Advanced
ORDINARY DIFFERENTIAL EQUATIONS

Advanced ORDINARY DIFFERENTIAL EQUATIONS

Athanassios G. Kartsatos
University of South Florida

Mariner Publishing Company, Inc.
Tampa, Florida

Copyright © 1980 by Mariner Publishing Company, Inc.

All Rights Reserved
Library of Congress Cataloging in Publication Data

Kartsatos, Athanassios G
   Advanced ordinary differential equations.

   Bibliography: p.
   Includes index.
   1. Differential equations. I. Title.
QA372.K26   515.3'52   80-10881

ISBN 0-936166-02-9

This book was typeset in Century Textbook
  by Mariner Publishing Company, Inc.

John W. Hobbes was the production manager.

Printed by Hillsboro Printing and Engraving
  Company and bound by Rose Printing Company

To My Parents:
Andromachi and Yorgos

PREFACE

This book is based on the author's lecture notes for advanced courses in ordinary differential equations at the University of South Florida. It is specially designed for a one-semester or a two-quarter course in advanced differential equations. Although written primarily for graduate or advanced undergraduate students of mathematics, this book should certainly prove to be quite useful also to physicists, engineers, and other scientists who are interested in various stability, asymptotic behavior, and boundary value problems concerning ordinary differential systems.

The book presupposes knowledge of a basic course in ordinary differential equations as well as a course in advanced calculus.

It is the intention of the author to exhibit and study the main problems of differential equations in the general setting of a differential system and to provide and utilize some of the main tools of functional analysis that may be employed for the solution of such problems. These functional analytic methods include the main fixed point theorems and the inverse function theorem in Banach spaces. Whenever such abstract methods are used in the book, the underlying space is a Banach space of "continuous" functions. This is by no means the only case of interest in applications, but such Banach spaces have enough structure to allow for the use of all these infinite dimesional techniques. The interested student will be able, after reading this book, to proceed further, and in a natural way, in the study of the application of functional analysis to differential equations in other suitable spaces. Such a study might require good knowledge of the Lebesgue measure and integration theory, which is not needed in this book.

Chapters One and Two are of introductory nature. Chapter One contains some basic properties of Banach spaces and bounded linear operators on such spaces. Here, the author has in mind that this might be the first time the student is exposed to infinite dimensional normed spaces. That is why this chapter contains only material that is absolutely necessary in later developments, as well as exercises and examples that enable the student to further understand the various theoretical concepts and results.

The three basic fixed point theorems and the inverse function theorem are the subject of Chapter Two. This is where the student actually realizes that solutions to various functional equations can be found to exist by the use of simple but powerful techniques of functional analysis. The applicability of these techniques involves the choice of the Banach space in which the problem is to be solved. It also involves the choice of the fixed point theorem to be used or the possibility of using the inverse function theorem. These considerations are reflected in some of the relevant exercises at the end of the chapter.

The first application of the Schauder-Tychonov theorem is given in the proof of Theorem 3.1. This theorem is the classical result of Peano concerning the local existence of solutions under mere continuity assumptions on the function involved. The Picard-Lindelof theorem is proved, independently of Peano's theorem, by using the usual Picard iterates as in the scalar case. Chapter Three also contains several results concerning the extendability of solutions to differential systems. In particular, it contains the basic existence and uniqueness theory for linear systems.

The stability of linear and perturbed linear systems is studied in Chapter Four. Necessary and sufficient conditions are given for a linear system to possess one of the five main stability properties. The conditions are stated in terms of fundamental matrices, or the measure of the coefficient matrix. Perturbations are then studied for such a system that are "small" enough to maintain its stability properties.

Chapter Five is devoted to the study of Lyapunov functions in connection with differential systems. The basic relationship between a differential system and corresponding scalar equations via Lyapunov functions is well exhibited and is applied to problems of continuation and stability.

Boundary value problems are contained in Chapter Six. Some fundamental results are given concerning periodic solutions of linear systems. A thorough treatment of more general boundary value problems for linear and perturbed linear systems is also included. The boundary conditions are linear throughout this chapter. It should be noted, however, that most of the results of this chapter are about the so-called "non-resonance" case. In this case the corresponding homogeneous problems admit only the zero solution. The resonance case is considered to be beyond the scope of this book. All the boundary value problems for perturbed linear systems are handled by use of fixed point theorems.

Chapter Seven contains a body of knowledge that is not usually contained in an organized way in other books at this level. The author's intention here is to introduce the student to the concept of monotonicity and its impact on the modern theory of ordinary differential equations. Several results are given, utilizing monotonicity assumptions, concerning the stability properties of perturbed linear systems. Two results on the existence of periodic solutions and a boundary value problem on an infinite interval are also included. The author believes that this is a good starting point toward the modern theory of abstract differential equations in Hilbert and Banach spaces.

Heavy use of fixed point theorems is made also in Chapter Eight. Fixed point theorems are employed in order to prove the existence of solutions of perturbed linear systems on the entire real line. Then conditions are given that ensure the periodicity or almost periodicity of such solutions. Fixed point theorems are also used in order to obtain stable solutions to quasilinear systems by using the information concerning the stability properties of the corresponding linear systems. Finally, applications of the inverse function theorem are given for boundary value problems of perturbed linear systems with linear boundary conditions. Further developments are also considered in cases where the boundary conditions are nonlinear but "Frechet differentiable." Then the Schauder-Tychonov theorem is employed again to extend these last generalizations to problems that contain also perturbations which are not Frechet differentiable.

The exercises at the end of each chapter are of various degrees of difficulty, and several among them are actually basic theoretical results.

The bibliography contains references to most of the existing books and some of the main papers related to this treatise.

The author would like to thank Professors Steve R. Bernfeld, William E. Fitzgibbon, III, John R. Graef, James R. Ward, Jr., Glenn F. Webb, and Jacob Wolfowitz, who took the time to go over the original manuscript and were kind enough to point out various changes for its improvement. Thanks are also due to the students in the author's classes for their help and their patience concerning the various versions of the manuscript. In particular, the author is indebted to Victor Dannon and Alan and Jessica Craig who read the manuscript several times over and located most of those hard-to-find flaws. To Janice, his wife, the author expresses his appreciation for her excellent typing of the manuscript, several times over, as well as her patience and her continuous encouragement and support.

A. G. Kartsatos
Tampa, Florida

CONTENTS

CHAPTER ONE   Banach Spaces 1
   1. The Concept of a Banach Space; the Space R^n 1
   2. Linear Operators 6
   3. Examples of Banach Spaces and Linear Operators 12
      Exercises 15

**CHAPTER TWO   Fixed Point Theorems in Banach Spaces;
The Inverse Function Theorem** 19
   1. The Contraction Principle 19
   2. The Schauder-Tychonov Theorem 22
   3. The Leray-Schauder Theorem 27
   4. The Inverse Function Theorem 28
      Exercises 34

**CHAPTER THREE   Existence, Uniqueness, and Continuation for
Differential Systems; Basic Theory of Linear Systems** 39
   1. Existence, Uniqueness 40
   2. Continuation 44
   3. Linear Systems 48
      Exercises 53

**CHAPTER FOUR   Stability of Linear Systems; Perturbed Linear
Systems** 59
   1. Definitions of Stability 59
   2. Linear Systems 60
   3. The Measure of a Matrix; Further Stability Criteria 65
   4. Perturbed Linear Systems 70
      Exercises 75

**CHAPTER FIVE   Lyapunov Functions in the Theory of Differential
Systems; the Comparison Principle** 79
   1. Lyapunov Functions 80
   2. Maximal and Minimal Solutions. The Comparison Principle .. 82
   3. Existence on R_+ 87
   4. The Comparison Principle and Stability 90
      Exercises 92

CHAPTER SIX Boundary Value Problems on Finite and Infinite Intervals ... 99
1. Linear Systems. Finite Intervals ... 100
2. Periodic Solutions of Linear Systems ... 101
3. Dependence of x(t) on A, U ... 105
4. Perturbed Linear Systems ... 109
5. Problems on Infinite Intervals ... 112
   Exercises ... 118

CHAPTER SEVEN Monotonicity in R^n and Differential Systems ... 123
1. A More General Inner Product for R^n ... 124
2. Stability of Differential Systems ... 126
3. Stability Regions ... 131
4. Periodic Solutions ... 135
5. Boundary Value Problems on Infinite Intervals ... 137
   Exercises ... 139

CHAPTER EIGHT Bounded Solutions on R; Quasilinear Systems; Applications of the Inverse Function Theorem ... 143
1. Exponential Dichotomies ... 144
2. Bounded Solutions on R ... 148
3. Quasilinear Systems ... 156
4. Applications of the Inverse Function Theorem ... 163
   Exercises ... 175

BIBLIOGRAPHY ... 181

INDEX ... 183

CHAPTER ONE

BANACH SPACES

In this chapter we develop the machinery that is needed throughout this book. We first introduce the concept of a real Banach space. This concept is of particular importance in the field of differential equations. Many problems in differential equations can actually be reduced to finding a solution of an equation of the form $Tx = y$. Here T is a certain operator mapping a subset of a Banach space X into another Banach space Y, and y is a fixed known element of Y. We next establish some fundamental properties of the Euclidean space of dimension n, as well as real $n \times n$ matrices. Then we introduce the concept of a bounded linear operator mapping a Banach space into another Banach space. We conclude this chapter by providing some examples of Banach spaces of continuous functions and bounded linear operators on such spaces.

1. The Concept of a Real Banach Space; the Space R^n

In what follows, the symbol "$x \in M$", or "$M \ni x$", means that x is an element of the set M. By the symbol "$A \subset B$" we mean that the set A is a subset of the set B. The symbol $f: A \to B$ means that the function f is defined on the set A and has values in the set B. By $\overline{M}$ we denote the closure of the set M. We use the symbol $\{x_n\}_{n=1}^{\infty} \subset M$ to denote the fact that the sequence $\{x_n\}_{n=1}^{\infty}$ has all of its terms in the set M.

We denote by R the real line, and by R_+, R_- the intervals $[0, \infty)$, $(-\infty, 0]$, respectively. The interval $[a, b] \subset R$ will always be finite; that is, $-\infty < a < b < +\infty$. By a "subspace" of the vector space X we mean a subset M of X that is itself a vector space with the same operations. The abbreviation "w.r.t." means "with respect to."

Definition 1.1 Let X be a vector space over R. Let $\|\cdot\|: X \to R_+$ have the following properties:

(i) $\|x\| = 0$ if and only if $x = 0$
(ii) $\|\alpha x\| = |\alpha|\|x\|$ for every $\alpha \in R$, $x \in X$
(iii) $\|x + y\| \leq \|x\| + \|y\|$ for every $x, y \in X$

Then the function $\|\cdot\|$ is called a "norm of X" and X is called a "real normed space."

The map $d(x, y) = \|x - y\|$ is a distance function on X. Thus (X, d) is a metric space. In what follows, the topology of a real normed space is assumed to be the one induced by its distance function — the "norm topology." We also use the term "normed space" instead of "real normed space." Without further mention, the symbol $\|\cdot\|$ always denotes the norm of the space that is being used.

We have Definitions 1.2, 1.3, and 1.4 concerning convergence in a normed space.

Definition 1.2 Let X be a normed space. The sequence $\{x_n\}_{n=1}^\infty \subset X$ "converges to $x \in X$" if the numerical sequence $\|x_n - x\|$ converges to zero as $n \to \infty$. We sometimes denote this by $x_n \to x$.

Definition 1.3 A sequence $\{x_n\}_{n=1}^\infty \subset X$, X a normed space, is said to be a "Cauchy sequence" if

$$\lim_{m,n \to \infty} \|x_m - x_n\| = 0$$

Definition 1.4 A normed space X is said to be "complete" if every Cauchy sequence in X converges to some element of X. A complete normed space is called a "Banach space."

The Euclidean space of dimension n is denoted by R^n. We let $R = R^1$. Unless otherwise specified, the vectors in R^n are assumed to be column vectors; that is vectors

$$x = \begin{bmatrix} x_1 \\ \vdots \\ x_n \end{bmatrix}$$

where $x_i \in R$, $i = 1, 2, \ldots, n$. Sometimes we also use the notation $x = (x_1, x_2, \ldots, x_n)$. The basis of R^n is always the set $\{e_1, e_2, \ldots, e_n\}$ where e_i has ith coordinate 1 and the rest 0.

Three different norms of R^n are given by Example 1.5.

Example 1.5 The space R^n is a Banach space if it is associated with any one of the following norms:

$$\|x\|_1 = (x_1^2 + \cdots + x_n^2)^{1/2}$$

$$\|x\|_2 = \max\{|x_1|, \ldots, |x_n|\}$$

$$\|x\|_3 = |x_1| + \cdots + |x_n|$$

Unless otherwise specified, R^n is always assumed to be associated with the first norm above, which is called the "Euclidean norm."

Definition 1.6 Let X be a normed space. Two norms $\|\cdot\|_a$, $\|\cdot\|_b$ of X are said to be "equivalent" if there exist positive constants m, M such that

$$m\|x\|_a \leq \|x\|_b \leq M\|x\|_a$$

for every $x \in X$.

The following theorem shows that any two norms of R^n are equivalent.

Theorem 1.7 If $\|\cdot\|_a$, $\|\cdot\|_b$ are two norms of R^n, then they are equivalent.

Proof. We recall that $\{e_1, e_2, \ldots, e_n\}$ is the natural basis of R^n; that is, e_i is the vector with ith coordinate 1 and all the rest 0. Let x be a vector in R^n. Then we have

$$x = \sum_{i=1}^{n} x_i e_i \tag{1.1}$$

Taking a-norms of both sides of (1.1) we get

$$\|x\|_a \leq \sum_{i=1}^{n} |x_i| \|e_i\|_a \tag{1.2}$$

This inequality implies

$$\|x\|_a \leq M\|x\|_1 \tag{1.3}$$

where $\|\cdot\|_1$ is the Euclidean norm and

$$M = \left[\sum_{i=1}^{n} \|e_i\|_a^2\right]^{1/2}$$

Here we have used the Cauchy-Schwarz inequality (see also Theorem 1.9 below).

It follows that for any $x, y \in R^n$ we have

$$|\|x\|_a - \|y\|_a| \le \|x - y\|_a \le M\|x - y\|_1 \qquad (1.4)$$

The first inequality in (1.4) is given as an exercise (see Exercises 1.1). From (1.4) we conclude that the function $\|x\|_a = f(x)$ is continuous on R^n w.r.t. the Euclidean norm. Since the sphere

$$S = \{u \in R^n;\ \|u\|_1 = 1\}$$

is compact, the function f attains its minimum m at some point $x_0 \in S$. Consequently, for every $u \in S$ we have $\|u\|_a \ge m$. Now let $x \in R^n$ be given with $x \ne 0$. Then $x/\|x\|_1 \in S$ and

$$\left\|\frac{x}{\|x\|_1}\right\|_a \ge m$$

which gives

$$\|x\|_a \ge m\|x\|_1 \qquad (1.5)$$

Since Equation (1.5) holds also for $x = 0$, we have that (1.5) is true for all $x \in R^n$. Inequalities (1.3) and (1.5) show that every norm of R^n is equivalent to the Euclidean norm. This proves our assertion.

The following definition concerns itself with the "inner product" in R^n. Theorem 1.9 contains the fundamental properties of the inner product.

Definition 1.8 The space R^n is associated with the inner product $\langle \cdot, \cdot \rangle$ defined as follows:

$$\langle x, y \rangle = \sum_{i=1}^{n} x_i y_i$$

Here $x = (x_1, \ldots, x_n)$, $y = (y_1, \ldots, y_n)$.

Theorem 1.9 For every $\alpha, \beta \in R$, $x, y, z \in R^n$ and for the Euclidean norm we have

(a) $\langle x, y \rangle = \langle y, x \rangle$
(b) $\langle x, \alpha y + \beta z \rangle = \alpha \langle x, y \rangle + \beta \langle x, z \rangle$
(c) $\|x\|^2 = \langle x, x \rangle \ge 0$
(d) $\langle x, x \rangle = 0$ if and only if $x = 0$
(e) $|\langle x, y \rangle| \le \|x\|\|y\|$ (Cauchy-Schwarz inequality)
(f) $\langle Ax, y \rangle = \langle x, A^T y \rangle$, where A^T denotes the transpose of the $n \times n$ real matrix A.

The proof is left as an exercise (see Exercise 1.13).

From elementary linear algebra we recall the following definitions, theorems, and auxiliary facts. We denote by C the complex plane and by M_n the real vector space of all real $n \times n$ matrices. For $A \in M_n$ we write $A = [a_{ij}]$, $i, j = 1, 2, \ldots, n$ or simply $A = [a_{ij}]$.

Definition 1.10 Two vectors $x, y \in R^n$ are called "orthogonal" if $<x, y> = 0$. A finite set $U = \{x_1, x_2, \ldots, x_n\} \in R^n$ is called "orthonormal" if any two vectors in U are orthogonal and $\|x_i\| = 1$, $i = 1, 2, \ldots, n$.

Definition 1.11 The number $\lambda \in C$ is called an "eigenvalue" of the matrix $A \in M_n$ if

$$|A - \lambda I| = 0$$

where $|\cdot|$ denotes determinant and I the identity matrix in M_n. Obviously, if λ is an eigenvalue of A, then the equation $(A - \lambda I)x = 0$ has at least one nonzero solution in C^n. Such a solution is called an "eigenvector" of A.

Theorem 1.12 A symmetric matrix $A \in M_n$ $(A^T = A)$ has only real eigenvalues. Moreover, A has a set of n linearly independent eigenvectors in R^n which is orthonormal.

Definition 1.13 A symmetric matrix $A \in M_n$ is said to be "positive definite" if

$$<Ax, x> \, > 0$$

for any $x \in R^n$ with $x \neq 0$.

Assume now that A is a symmetric matrix in M_n. Then the continuous function

$$\phi(u) = <Au, u>$$

attains its maximum λ_M and its minimum λ_m on the unit sphere

$$S = \{u \in R^n; \|u\| = 1\}.$$

Let $\lambda_M = <Au_0, u_0>$, $\lambda_m = <Av_0, v_0>$ for some $u_0, v_0 \in S$. We consider the function $g(x) = x_1^2 + x_2^2 + \cdots + x_n^2 - 1$. It is easy to see that $\nabla \phi(x) = 2Ax$ and $\nabla g(x) = 2x$. Since S is the set of all points x such that $g(x) = 0$ and $\nabla g(x) \neq 0$, it follows from a well known theorem of Advanced Calculus (see, for example, Edwards [13, p.108]) that there exist real numbers λ, μ such that

$$\nabla \phi(u_0) = \lambda \nabla g(u_0), \quad \nabla \phi(v_0) = \mu \nabla g(v_0) \tag{1.6}$$

or $Au_0 = \lambda u_0$ and $Av_0 = \mu v_0$. Since $<Au_0, u_0> = \lambda$, $<Av_0, v_0> = \mu$, we have $\lambda = \lambda_M$ and $\mu = \lambda_m$.

Theorem 1.14 If λ_m, λ_M are the smallest and largest eigenvalues of a symmetric matrix $A \in M_n$, respectively, then we have

$$\lambda_m = \min_{\|u\|=1} <Au, u> \tag{1.7}$$

$$\lambda_M = \max_{\|u\|=1} <Au, u> \tag{1.8}$$

If A is positive definite, then all the eigenvalues of A are positive.

Definition 1.15 A matrix $P \in M_n$ is called a "projection matrix" if $P^2 = P$.

It is easy to see that if P is a projection matrix, then $I - P$ is also a projection matrix.

2. Linear Operators

In what follows, an "operator" is simply a function mapping a subset of a Banach space into another Banach space. In this section we obtain some elementary information concerning bounded linear operators. We also provide three norms for the space M_n which correspond to the norms given for R^n in Example 1.5. In particular, we recall some facts concerning linear operators mapping R^n into itself.

Definition 1.16 Let X, Y be two normed spaces, and let V be a subset of X. Then an operator $T: V \to Y$ is "continuous at $x_0 \in V$" if $Tx_n \to Tx_0$ for any sequence $\{x_n\}_{n=1}^\infty \subset V$ with $x_n \to x_0$. T is "continuous on V" if it is continuous at each $x_0 \in V$.

The following definition concerns itself with an important class of operators — the linear operators.

Definition 1.17 Let X, Y be two normed spaces and V a subspace of X. Then $T: V \to Y$ is called "linear" if for every $\alpha, \beta \in R$, $x, y \in V$ we have

$$T(\alpha x + \beta y) = \alpha Tx + \beta Ty$$

We often omit the parentheses in $T(x)$ for operators that are considered in the sequel.

Definition 1.18 Let X, Y be two normed spaces. A linear operator $T: X \to Y$ is called "bounded" if there exists a constant $K \geq 0$ such that $\|Tx\| \leq K\|x\|$ for every $x \in X$. If T is bounded, then the number

$$\|T\| = \sup_{\|x\| \leq 1} \|Tx\|$$

is called the "norm of T."

Unless otherwise specified, the symbol $\|\cdot\|$ denotes the norm of all Banach spaces and bounded linear operators under consideration.

Theorem 1.19 below characterizes the continuous linear operators on Banach spaces.

Theorem 1.19 A linear operator $T: X \to Y$, with X, Y normed spaces, is continuous on X if and only if it is bounded.

Proof. Sufficiency. From the inequality

$$\|Tx\| \leq K\|x\|, \quad x \in X \tag{1.9}$$

it follows immediately that

$$\|Tx - Tx_0\| \leq K\|x - x_0\|$$

for any $x_0, x \in X$. Thus, if $x \to x_0$, then $Tx \to Tx_0$.

Necessity. Suppose that T is continuous on X. We show that

$$K_0 = \sup_{\|x\|=1} \|Tx\| < +\infty$$

In fact, let $K_0 = +\infty$. Then there exists a sequence $\{x_n\}_{n=1}^{\infty} \subset X$ such that $\|x_n\| = 1$ and $0 < \lambda_n = \|Tx^n\| \to \infty$. Let $\bar{x}_n = x_n/\lambda_n$. Then, obviously, $\|\bar{x}_n\| = (1/\lambda_n)\|x_n\| \to 0$ and $\|T\bar{x}_n\| = 1$; that is, a contradiction to the continuity of T. Therefore, $K_0 < +\infty$. Let $x \neq 0$ be a vector in X. Then $\bar{x} = x/\|x\|$ satisfies $\|\bar{x}\| = 1$. Thus, $\|T\bar{x}\| = \|Tx\|/\|x\|$ and $\|T\bar{x}\| \leq K_0$. Consequently, we have

$$\|Tx\| \leq K_0 \|x\| \tag{1.10}$$

Since (1.10) also holds for $x = 0$, we have shown (1.9) with $K = K_0$.

Theorem 1.20 Let X, Y be two normed spaces. Let $T: X \to Y$ be a bounded linear operator. Then

$$\|T\| = \sup_{\|x\|=1} \|Tx\|$$

Proof. Obviously,

$$\sup_{\|x\|=1} \|Tx\| \leq \sup_{\|x\|\leq 1} \|Tx\|$$

Let $x \in X$ be such that $\|x\| \leq 1$ and $x \neq 0$. Then

$$\|Tx\| = \|x\| \|T(x/\|x\|)\| \leq \sup_{\|x\|=1} \|Tx\|$$

Let X, Y be two normed spaces. The space of all bounded linear operators $T: X \to Y$ is a vector space under the obvious definitions of addition and multiplication by scalars (reals). This space becomes a normed space if it is associated with the norm of Definition 1.18. For $X = Y = R^n$ we have

Example 1.21 Let A be a matrix in M_n. Consider the operator $T: R^n \to R^n$ defined by

$$Tx = Ax, \; x \in R^n$$

Then T is a linear operator. Now let $A = [a_{ij}]$, $i, j = 1, 2, \ldots, n$. Then R^n associated with the three norms of Example 1.5 induces three norms on T according to Table 1.1.

Table 1.1

$\|x\|$	$\|T\|$
$\|x\|_1$	$\sqrt{\lambda}$ (λ is the largest eigenvalue of $A^T A$)
$\|x\|_2$	$\sup_i \sum_j |a_{ij}|$
$\|x\|_3$	$\sup_j \sum_i |a_{ij}|$

We prove the first assertion in the above table. The others are left as an exercise (see Exercise 1.11).

Theorem 1.22 Let R^n be associated with the Euclidean norm. Let T be the linear operator of Example 1.21. Then

$$\|T\| = \sqrt{\lambda}$$

where λ is the largest eigenvalue of $A^T A$.

Proof. We assume first that A is symmetric and that λ_1 is an eigenvalue of A with largest absolute value $|\lambda_1|$. In this case $\|T\| = |\lambda_1|$. In fact, let $\lambda_1, \lambda_2, \ldots, \lambda_n$ be the (real) eigenvalues of A with n corresponding real eigenvectors x_i, $i = 1, 2, \ldots, n$, that form an orthonormal set. Then these vectors are linearly independent.

In fact, if

$$c_1 x_1 + c_2 x_2 + \cdots + c_n x_n = 0$$

with c_i, $i = 1, 2, \ldots, n$, real constants, then

$$c_1 <x_1, x_1> + c_2 <x_2, x_1> + \cdots + c_n <x_n, x_1> = 0$$

showing that $c_1 = 0$. Similarly, $c_i = 0$, $i = 1, 2, \ldots, n$. Thus, the set $\{x_1, \ldots, x_n\}$ is a basis for R^n. Let $x \in R^n$ be given and let $\{c_1, c_2, \ldots, c_n\}$ be a set of real constants with

$$x = c_1 x_1 + c_2 x_2 + \cdots + c_n x_n$$

Then we have

$$Tx = T(c_1 x_1 + c_2 x_2 + \cdots + c_n x_n)$$
$$= c_1 \lambda_1 x_1 + c_2 \lambda_2 x_2 + \cdots + c_n \lambda_n x_n$$

and

$$\|Tx\|^2 = <Tx, Tx> = |c_1 \lambda_1|^2 + |c_2 \lambda_2|^2 + \cdots + |c_n \lambda_n|^2$$
$$\leq \lambda_1^2 (|c_1|^2 + \cdots + |c_n|^2) = \lambda_1^2 \|x\|^2$$

It follows that $\|Tx\| \leq |\lambda_1| \|x\|$ for every $x \in R^n$. Since $\|Tx_1\| = |\lambda_1|$, we obtain that $\|T\| = |\lambda_1|$. We also have the following characterization for $\|T\|$:

$$\|T\| = \sup_{\|x\|=1} |<Ax, x>| = \sup_{\|x\|=1} |<Tx, x>| \tag{1.11}$$

In fact, let $x \in R^n$ be given with $\|x\| = 1$. Then

$$|<Tx, x>| \le \|Tx\|\|x\| \le \|T\|\|x\| = \|T\| = |\lambda_1|$$

Moreover, we have $|<Tx, x>| = |\lambda_1|$ for $x = x_1$. Consequently,

$$\sup_{\|x\|=1} |<Tx, x>| = |\lambda_1|$$

proving (1.11). Now let $A \in M_n$ be arbitrary. We have

$$\|T\|^2 = \sup_{\|x\|=1} \|Tx\|^2 = \sup_{\|x\|=1} <Tx, Tx> = \sup_{\|x\|=1} <Ax, Ax>$$

$$= \sup_{\|x\|=1} <A^T Ax, x> = \lambda$$

where $\lambda = |\lambda|$ is the largest eigenvalue of $A^T A$.

In the following discussion, we identify A and T in Example 1.21 and we assume (unless otherwise specified) that $A \in M_n$ has norm $\|A\| = \|T\| = \sqrt{\lambda}$, as in Theorem 1.22. It is easy to see that M_n is a Banach space with any one of the norms given by Table 1.1.

Now let $P \in M_n$ be a projection matrix. We also denote by P the linear operator defined by P as above. Then

$$R^n = PR^n \oplus (I - P)R^n$$

that is, R^n is the direct sum of the subspaces PR^n, $(I - P)R^n$. The equation $R^n = M \oplus N$, with M, N subspaces of R^n, means that every $x \in R^n$ can be written in a unique way as $y + z$, where $y \in M$, $z \in N$. Let us first show that $PR^n \cap (I - P)R^n = \{0\}$. Assume $x \in PR^n \cap (I - P)R^n$. Then there exist y, z in R^n such that $x = Py = (I - P)z$. This implies

$$Px = P^2 y = Py = P(I - P)z = (P - P^2)z = (P - P)z = 0$$

Thus, $x = Py = 0$. Assume now that $x = y + z = y_1 + z_1$ with $y, y_1 \in PR^n$ and $z, z_1 \in (I - P)R^n$. Then

$$PR^n \ni y - y_1 = z_1 - z \in (I - P)R^n$$

implies that $y - y_1 = z_1 - z = 0$. This says that $y = y_1$ and $z = z_1$.

We summarize the above in the following theorem.

Theorem 1.23 Let $P \in M_n$ be a projection matrix. Then $R^n = PR^n \oplus (I-P)R^n$.

Now we give a meaning to the symbol e^A, where A is a matrix in M_n. We consider the series

$$I + \sum_{m=1}^{\infty} (A^m/m!) \qquad (1.12)$$

Since M_n is complete, the convergence of the series in (1.12) will be shown if we prove that the sequence of partial sums $\{S_m\}_{m=1}^{\infty}$ with

$$S_m = I + \sum_{k=1}^{m} (A^k/k!)$$

is Cauchy. To this end, we observe that

$$\|S_{\bar{m}} - S_m\| \le \sum_{i=m+1}^{\bar{m}} (\|A\|^i/i!)$$

for every $\bar{m} > m$ and that

$$\sum_{i=0}^{\infty} (\|A\|^i/i!) = e^{\|A\|}$$

It follows that $\{S_m\}$ is a Cauchy sequence. We denote its limit by e^A. Then it is easy to see that if $A, B \in M_n$ commute, then $e^{A+B} = e^A e^B$. From this equality we obtain

$$e^A \cdot e^{-A} = e^{-A} \cdot e^A = e^0 = I$$

Thus, if A^{-1} denotes the inverse of the matrix A, we obtain

$$[e^A]^{-1} = e^{-A} \quad \text{for every } A \in M_n$$

We now recall some elementary facts from Advanced Calculus. Let J be a real interval. Let $f: J \to R^n$ be such that $f(t) = (f_1(t), \ldots, f_n(t))$, $t \in J$. Then f is differentiable at $t_0 \in J$ if and only if each function f_i is differentiable at t_0. We have

$$f'(t_0) = (f_1'(t_0), \ldots, f_n'(t_0))$$

Similarly, f is (Riemann) integrable on $[a, b] \subset J$ if and only if each function f_i is (Riemann) integrable on $[a, b]$. We have

$$\int_a^b f(t)dt = (\int_a^b f_1(t)dt, \ldots, \int_a^b f_n(t)dt)$$

An analogous situation exists for functions $A: J \to M_n$ with $A(t) = [a_{ij}(t)]$, $i, j = 1, 2, \ldots, n$. Exercises 1.6, 1.7 and 1.23 contain several differentiation and integration properties of functions $f(t)$, $A(t)$. All these properties are assumed to hold without further mention.

3. Examples of Banach Spaces and Linear Operators

If J is any interval of R, finite or infinite, and $f: J \to R^n$ is a bounded function, we set

$$\|f\|_\infty = \sup_{t \in J} \|f(t)\|$$

Although the symbol $\|\cdot\|_\infty$ does not refer to the interval J, no confusion can be created this way because the supremum is taken over the entire interval of definition of the function f under consideration. We also let $C_n(J)$ denote the vector space of all bounded and continuous functions defined on J and having values in R^n. If the interval is denoted by two points a, b, with $|a|, |b| \leq +\infty$ and $a < b$, then we drop the parentheses in $C_n(J)$. Thus we have $C_n[a, b)$, $C_n(a, b)$, etc.

Example 1.24 The space $C_n(J)$, with J any interval of R, is a Banach space with norm

$$\|f\|_\infty = \sup_{t \in J} \|f(t)\|$$

Example 1.25 Let T be a positive number and let $P_n(T)$ be the space of all continuous T-periodic functions with values in R^n, that is,

$$P_n(T) = \{x \in C_n(R);\ x(t + T) = x(t),\ t \in R\}$$

Then $P_n(T)$ is a Banach space with norm

$$\|f\|_\infty = \max_{t \in [0,T]} \|f(t)\|$$

The space $P_n(T)$ can be identified with the space

$$\{x \in C_n[0, T];\ x(0) = x(T)\} \tag{1.13}$$

In fact, every function $f \in P_n(T)$ determines uniquely a function in (1.13) — its restriction $\bar{f}$ on the interval $[0, T]$. Conversely, every function $\bar{u}$ in the space (1.13) can be extended uniquely to a function $u \in P_n(T)$ in an obvious way. The correspondence $f \to \bar{f}$ is an isomorphism onto and such that

$$\|f\| = \|\bar{f}\|$$

The space $P_n(T)$ is thus a closed subspace of the Banach space $C_n[0, T]$.

Example 1.26 The space C_n^l of all functions $f \in C_n[0, \infty)$ with finite limits as $t \to \infty$ is a closed subspace of $C_n[0, \infty)$. Thus, it is a Banach space with norm

$$\|f\|_\infty = \sup_{t \in R_+} \|f(t)\|$$

C_n^l contains the space C_n^0 of all functions $f \in C_n^l$ such that

$$\lim_{t \to \infty} f(t) = 0$$

C_n^0 is also a closed subspace of $C_n[0, \infty)$, and thus it is a Banach space with the sup-norm. The interval R_+ can be replaced in this example by any interval $[t_0, \infty)$, for some $t_0 \in R$.

Definition 1.27 A function $f \in C_n(R)$ is called "almost periodic" if for every $\epsilon > 0$ there exists a number $L(\epsilon) > 0$ such that every interval of length $L(\epsilon)$ contains at least one number τ such that

$$\|f(t + \tau) - f(t)\| < \epsilon$$

for every $t \in R$.

The following theorem gives a characterization of almost periodicity.

Theorem 1.28 A function $f \in C_n(R)$ is almost periodic if and only if every sequence $\{f(t + \tau_m)\}_{m=1}^\infty$ of translates of f contains a uniformly convergent subsequence. Here $\{\tau_m\}_{m=1}^\infty$ is any sequence of constants in R.

Example 1.29 The space AP_n of all almost periodic functions $f \in C_n(R)$ is a closed subspace of $C_n(R)$, and thus a Banach space with norm

$$\|f\|_\infty = \sup_{t \in R} \|f(t)\|$$

Several properties of almost periodic functions can be found in some exercises at the end of this chapter.

Example 1.30 Let $J \subset R$ be an interval. We define the space $C_n^k(J)$, $k = 1, 2, \ldots$, as follows:

$$C_n^k(J) = \{f \in C_n(J); \ f^{(k)} \in C_n(J)\}$$

This space is a Banach space with norm

$$\|f\|_* = \sum_{i=0}^{k} \sup_{t \in J} \|f^{(i)}(t)\| = \sum_{i=0}^{k} \|f^{(i)}\|_\infty$$

It should be noted here that derivatives at finite left (right) endpoints of intervals of R are right (left) derivatives. We also let $f^{(0)} = f$.

In the following examples we provide some bounded linear operators on Banach spaces. An operator T from a Banach space into R is called a functional. A bounded linear functional is given in Example 1.31.

Example 1.31 Consider the operator $T: C_1[a, b] \to R$ with

$$Tx = \sum_{i=1}^{n} c_i x(t_i)$$

where $c_1, c_2, \ldots, c_n$ are fixed real constants and $t_1, t_2, \ldots, t_n$ are fixed points in $[a, b]$. Then T is a bounded linear operator with norm

$$\|T\| = \sum_{i=1}^{n} |c_i| \tag{1.14}$$

Obviously,

$$|Tx| \leq M \|x\|_\infty \quad \text{for every } x \in C_1[a, b]$$

where M is the number in the right hand side of (1.14). Now it is easy to find some $\bar{x} \in C_1[a, b]$ such that $\|\bar{x}\|_\infty = 1$ and $|T\bar{x}| = M$. This shows that (1.14) holds.

Example 1.32 Let T denote the operator which maps every function $x \in C_n[a, b]$ into the function

$$y(t) = \int_a^b K(t, s) x(s) ds$$

Here $K: [a, b] \times [a, b] \to M_n$ is a continuous function; that is, all the entries of K are continuous on $[a, b] \times [a, b]$. Then T is linear and

$$\|Tx\|_\infty \leq \max_{t \in [a,b]} \int_a^b \|K(t, s)\| ds \|x\|_\infty$$

Thus

$$\|T\| \leq \max_{t \in [a,b]} \int_a^b \|K(t, s)\| ds$$

Example 1.33 Consider the operator $T: C_2^2[a, b] \to C_2[a, b]$ with

$$(Tx)(t) = x''(t) + x(t)$$

Then $\|Tx\|_\infty \leq \|x\|_2$. Here $\|\cdot\|_2$ is the norm in $C_2^2[a, b]$. Let $\bar{x} \in C_2^2[a, b]$ be given by

$$\bar{x}(t) = \begin{bmatrix} 1 \\ 0 \end{bmatrix}, \quad t \in [a, b]$$

Then $\|\bar{x}\|_2 = 1$ and $\|T\bar{x}\|_\infty = 1$. Consequently, $\|T\| = 1$.

Example 1.34 Let the function $A: R_+ \to M_n$ be continuous and such that

$$\int_0^\infty \|A(t)\| dt < +\infty$$

Consider the operator $T: C_n(R_+) \to C_n^l$ defined as follows:

$$(Tx)(t) = L + \int_t^\infty A(s)x(s)ds$$

Here L is a fixed vector in R^n. It is easy to see that $(Tx)(t) \to L$ as $t \to \infty$. We also have

$$\|(Tx)(t)\| \leq \|L\| + \left\| \int_t^\infty A(s)x(s)ds \right\|$$

$$\leq \|L\| + \int_t^\infty \|A(s)\| \|x(s)\| ds$$

$$\leq \|L\| + \int_0^\infty \|A(s)\| ds \|x\|_\infty$$

If $L = 0$, we obtain

$$\|Tx\|_\infty \leq M \|x\|_\infty$$

with

$$M = \int_0^\infty \|A(t)\| dt$$

Consequently, for $L = 0$, T is a bounded linear operator with $\|T\| \leq M$.

EXERCISES

1.1. Let X be a normed space. Let $x, y \in X$ be given. Show that

$$|\|x\| - \|y\|| \leq \|x - y\|$$

1.2. For $f \in C_1[a, b]$, let
$$\|f\| = \left(\int_a^b |f(t)|^2 dt\right)^{1/2}$$
Show that $C_1[a, b]$ is not complete with this norm.

1.3. Show that the spaces $C_n^k[a, b]$, $C_n(R_+)$, $C_n^{l'}$, C_n^0 are Banach spaces with norms as in the examples of Section 3 above.

1.4. Let X, Y be Banach spaces. Let $T: X \to Y$ be a linear operator. Show that T is continuous at $x_0 \in X$ if and only if T is continuous at 0.

1.5. Let X, Y, Z be Banach spaces. Let $T: X \to Y$, $U: Y \to Z$ be bounded linear operators. Show that
$$\|UT\| \le \|U\| \|T\|$$
Here UT denotes composition; that is, $(UT)(x) = U(Tx)$, $x \in X$. Conclude that for any matrices $A, B \in M_n$ we have
$$\|AB\| \le \|A\| \|B\|$$

1.6. Let $f: [a, b] \to R^n$ be continuous. Prove the following inequality, and then state and prove an analogous inequality for M_n-valued functions.
$$\left\| \int_a^b f(t) dt \right\| \le \int_a^b \|f(t)\| dt$$

1.7. Assume that $x, y: [a, b] \to R^n$, $A, B: [a, b] \to M_n$ are differentiable at the point $t \in [a, b]$. Prove the following differentiation rules:
 (i) $[A(t) + B(t)]' = A'(t) + B'(t)$
 (ii) $[cA(t)]' = cA'(t)$, $c \in R$
 (iii) $[A(t)B(t)]' = A'(t)B(t) + A(t)B'(t)$
 (iv) $[A(t)x(t)]' = A'(t)x(t) + A(t)x'(t)$
 (v) $<x(t), y(t)>' = <x'(t), y(t)> + <x(t), y'(t)>$
 (vi) $[A^{-1}(t)]' = -A^{-1}(t)A'(t)A^{-1}(t)$ if A is nonsingular on $[a, b]$.

Show that the integration properties hold analogous to those of (i) and (ii).

1.8. Show that if $f: R^n \to R$ is a bounded linear functional, there exist constants $c_1, \ldots, c_n \in R$ such that
$$f(x) = c_1 x_1 + \cdots + c_n x_n$$
for every $x = (x_1, x_2, \ldots, x_n) \in R^n$.

1.9. Show that if x_0 is a given vector in R^n, then there exists a bounded linear functional $f: R^n \to R$ such that $f(x_0) = \|x_0\|$ and $\|f\| = 1$.

1.10. Let $x \in C_n^1[a, b]$ and $t_0, t_1 \in [a, b]$ be given. Show that there exists a number $\bar{t}$, properly between t_0 and t_1, such that
$$\|x(t_1) - x(t_0)\| \leq \|x'(\bar{t})\| |t_1 - t_0|$$
[Hint. Apply the Mean Value theorem to the function $t \to f(x(t))$, where $f: R^n \to R$ is a functional with
$$f(x(t_1) - x(t_0)) = \|x(t_1) - x(t_0)\|$$
and $\|f\| = 1$. See Exercise 1.9.]

1.11. Prove the last two cases of Example 1.21.

1.12. Let $T: X \to Y$ (X, Y Banach spaces) be a bounded linear operator mapping X onto Y. Show that the inverse operator $T^{-1}: Y \to X$ exists and is bounded if and only if there exists a positive constant m with the property:
$$\|Tx\| \geq m\|x\|, \quad x \in X$$

1.13. Prove Theorem 1.9. Then show that the matrix $A \in M_n$ is symmetric if and only if
$$<Ax, y> = <x, Ay>$$
for every $x, y \in R^n$.

1.14. Let $P \in M_n$ be a projection matrix. Show that
$$<Px, (I - P)y> = 0$$
for every $x, y \in R^n$ if and only if P is symmetric. [Hint. Use Exercise 1.13.]

1.15. Complete the proof of Example 1.31. Find a function $\bar{x}$ there with the desired properties.

1.16. Prove Theorem 1.28.

1.17. Show that every $f \in AP_n$ is uniformly continuous.

1.18. From Exercise 1.17 it follows that the function $f(t) = \sin t + \sin(qt)$ is almost periodic. Show that $f(t)$ is not T-periodic, for any number $T > 0$, if q is an irrational positive number.

1.19. Show that given two functions $f, g \in AP_n$ we have the following property: for every $\epsilon > 0$ there exists $L(\epsilon) > 0$ such that every interval of length $L(\epsilon)$ contains at least one number τ such that
$$\|f(t + \tau) - f(t)\| < \epsilon, \quad \|g(t + \tau) - g(t)\| < \epsilon$$

for every $t \in R$. Then show that AP_n is a Banach Space. [Hint. Consider the function
$$t \to \begin{bmatrix} f(t) \\ g(t) \end{bmatrix} \in R^{2n}$$
Show that this function is almost periodic by using the result of Theorem 1.28.]

1.20. Let M be a subspace of R^n. Show that there exists a subspace N of R^n such that $<x, y> = 0$ for $x \in M$, $y \in N$ and $R^n = M \oplus N$.

1.21. Consider the function $P: R^2 \to R^2$ which maps every point $(x_1, x_2) \in R^2$ into the point Q of the line $x_2 = 3x_1$ such that the distance between (x_1, x_2) and Q is a minimum. Show that P is a projection operator; that is, P represents a projection matrix in M_2.

1.22. Let $A: [a, b] \to M_n$ be a continuous function such that $A(t)$ is symmetric for every $t \in [a, b]$. Show that the largest and the smallest eigenvalues of $A(t)$ are continuous on $[a, b]$.

1.23. Given $A: [a, \infty) \to M_n$ continuous, we have
$$\int_a^t A(s)ds = \left[\int_a^t a_{ij}(s)ds \right], \quad t \in [a, \infty)$$
where $A(t) = [a_{ij}(t)]$, $t \in [a, b]$. Let
$$\int_a^\infty \|A(s)\| ds < +\infty$$
Show that
$$\lim_{t \to \infty} \int_a^t A(s)ds = \int_a^\infty A(s)ds$$
exists as a finite matrix and
$$\int_a^\infty A(s)ds = [\int_a^\infty a_{ij}(s)ds]$$

1.24. Show that if $C_2^1[0, 1]$ is associated with the sup—norm on $[0, 1]$, then it is not complete.

1.25. Let $C_2^1[0, 1]$ be associated with the sup—norm on $[0, 1]$. Let $T: C_2^1[0, 1] \to C_2[0, 1]$ be given by $(Tx)(t) = x(t) - 2x'(t)$, $t \in [0, 1]$. Show that T is not a bounded linear operator.

CHAPTER TWO

FIXED POINT THEOREMS IN BANACH SPACES; THE INVERSE FUNCTION THEOREM

In Chapter One we stated that a large number of problems in differential equations can be reduced to finding a solution x to an equation of the form $Tx = y$. The operator T maps a subset of a Banach space X into some other Banach space Y and y is a known element of Y. If $y = 0$ and $Tx = Ux - x$, for some other operator U, then the equation $Tx = y$ is equivalent to the equation $Ux = x$. Naturally, in order to solve $Ux = x$, we must assume that the range $R(U)$ and the domain $D(U)$ have points in common. Points x for which $Ux = x$ are called fixed points of the operator U.

In this chapter we state the main fixed point theorems that are most widely used in the field of differential equations. These are the Banach contraction principle, the Schauder-Tychonov theorem, and the Leray-Schauder theorem. We only prove the first theorem, because the proofs of the other two are beyond the scope of this book.

In the last section of this chapter we state and prove the inverse function theorem in Banach spaces. This theorem generalizes the well known inverse function theorem of Advanced Calculus.

1. The Contraction Principle

Theorem 2.1 (Banach contraction principle). Let x be a Banach space and M a nonempty closed subset of X. Let $T: M \to M$ be an operator such that there exists a constant $k \in [0,1)$ with the property

$$\|Tx - Ty\| \leq k\|x - y\| \quad \text{for every } x, y \in M \tag{2.1}$$

Then T has a unique fixed point in M.

Proof. Let $x_0 \in M$ be given with $T(x_0) \neq x_0$. Define the sequence $\{x_m\}_0^\infty$ as follows:

$$x_j = Tx_{j-1}, \quad j = 1, 2, \ldots \tag{2.2}$$

Then we have

$$\|x_{j+1} - x_j\| \leq k\|x_j - x_{j-1}\| \leq k^2\|x_{j-1} - x_{j-2}\|$$

$$\leq \ldots \leq k^j\|x_1 - x_0\|$$

for every $j \geq 1$. Thus, if $m > n \geq 1$, we obtain

$$\|x_m - x_n\| \leq \|x_m - x_{m-1}\| + \|x_{m-1} - x_{m-2}\| + \cdots + \|x_{n+1} - x_n\|$$

$$\leq k^{m-1}\|x_1 - x_0\| + k^{m-2}\|x_1 - x_0\| + \cdots + k^n\|x_1 - x_0\|$$

$$\leq k^n(1 + k + k^2 + \cdots + k^{m-n-1})\|x_1 - x_0\|$$

$$\leq [k^n/(1-k)]\|x_1 - x_0\|$$

Since $k^n \to 0$ as $n \to \infty$, it follows that $\{x_m\}_{m=0}^\infty$ is a Cauchy sequence. Since X is complete, there exists $\bar{x} \in X$ such that $x_m \to \bar{x}$. Obviously, $\bar{x} \in M$ because M is closed. Taking limits as $j \to \infty$ in (2.2), we obtain $\bar{x} = T\bar{x}$.

To show uniqueness, let y be another fixed point of T in M. Then

$$\|\bar{x} - y\| = \|T\bar{x} - Ty\| \leq k\|\bar{x} - y\|$$

which implies $\bar{x} = y$. This completes the proof of the theorem.

An operator $T: M \to X$, $M \subset X$, satisfying (2.1) on M is called a "contraction operator on M."

Example 2.2 Let X be a Banach space and let $T: X \to X$ be a bounded linear operator such that $\|T\| < 1$. Then T is a contraction operator on X.

Thus contractions can be easily obtained from Examples 1.32 and 1.34.

Example 2.3 Let $F: R_+ \times R^n \to R^n$ be continuous and such that

$$\|F(t, x_1) - F(t, x_2)\| \leq \lambda(t) \|x_1 - x_2\|$$

for every $t \in R_+$, $x_1, x_2 \in R^n$, where $\lambda: R_+ \to R_+$ is continuous and such that

$$L = \int_0^\infty \lambda(t) dt < +\infty$$

Assume further that

$$\int_0^\infty \|F(t, 0)\| dt < +\infty$$

Then the operator T with

$$(Tx)(t) = \int_t^\infty F(s, x(s)) ds, \quad t \in R_+$$

maps the space $C_n(R_+)$ into itself and is a contraction operator on $C_n(R_+)$ if $L < 1$. In fact, let $x, y, C_n(R_+)$ be given. Then we have

$$\|Tx\|_\infty \leq \int_0^\infty \|F(t, x(t))\| dt$$
$$\leq \int_0^\infty \|F(t, x(t)) - F(t, 0)\| dt + \int_0^\infty \|F(t, 0)\| dt$$
$$\leq \int_0^\infty \lambda(t) \|x(t)\| dt + \int_0^\infty \|F(t, 0)\| dt$$
$$\leq \int_0^\infty \lambda(t) dt \|x\|_\infty + \int_0^\infty \|F(t, 0)\|$$

which shows that $TC_n(R_+) \subset C_n(R_+)$. We also have

$$\|Tx - Tx\|_\infty \leq \int_0^\infty \|F(t, x(t)) - F(t, y(t))\| dt$$
$$\leq \int_0^\infty \lambda(t) \|x(t) - y(t)\| dt$$
$$\leq L \|x - y\|_\infty$$

It follows that, for $L < 1$, the equation $Tx = x$ has a unique solution $\bar{x}$ in $C_n(R_+)$. Thus, there is a unique $\bar{x} \in C_n(R_+)$ such that

$$\bar{x}(t) = \int_t^\infty F(s, \bar{x}(s))ds, \ t \in R_+$$

It is easily seen that under the above assumptions on F, L the equation

$$x(t) = f(t) + \int_t^\infty F(s, x(s))ds$$

also has a unique solution in $C_n(R_+)$ if f is a fixed function in $C_n(R_+)$. This solution belongs to C_n^1 if $f \in C_n^1$.

2. The Schauder-Tychonov Theorem

Before we state the Schauder-Tychonov theorem, we characterize the compact subsets of $C_n[a,b]$. This characterization, which is contained in Theorem 2.5, allows us to detect the relative compactness of the range of an operator defined on a subset of $C_n[a, b]$ and having values in $C_n[a, b]$. We define below the concept of a relatively compact, and a compact, set in a Banach space.

Definition 2.4 Let X be a Banach space. Then a subset M of X is said to be "compact" if every sequence $\{x_n\}_{n=1}^\infty$ in M contains a subsequence which converges to a vector in M. The set M is said to be "relatively compact" if every sequence $\{x_n\}_{n=1}^\infty$ from M contains a subsequence which converges to a vector in X.

It is obvious from this definition that M is relatively compact if and only if $\overline{M}$ (the closure of M in the norm of X) is compact. The following theorem characterizes the compact subsets of $C_n[a, b]$.

Theorem 2.5 Let M be a subset of $C_n[a,b]$. Then M is relatively compact if and only if

(i) There exists a constant K such that

$$\|f\|_\infty \leq K, f \in M$$

(ii) The set M is "equicontinuous;" that is, for every $\epsilon > 0$ there exists $\delta(\epsilon) > 0$ (depending only on ϵ) such that $\|f(t_1) - f(t_2)\| < \epsilon$ for all $t_1, t_2 \in [a, b]$ with $|t_1 - t_2| < \delta(\epsilon)$ and all $f \in M$.

The proof is based on Lemma 2.7. We start with Definition 2.6.

Definition 2.6 Let M be a subset of the Banach space X and let $\epsilon > 0$ be given. Then the set $M_1 \subset X$ is said to be an "ϵ-net of M" if for every point $x \in M$ there exists $y \in M_1$ such that $\|x - y\| < \epsilon$.

Lemma 2.7 Let M be a subset of a Banach space X. Then M is relatively compact if and only if for every $\epsilon > 0$ there exists a finite ϵ-net of M in X.

Proof. Necessity. Assume that M is relatively compact and that the condition in the statement of the lemma is not satisfied. Then there exists some $\epsilon_o > 0$ for which there is no finite ϵ_o-net of M. Choose $x_1 \in M$. Then $\{x_1\}$ is not an ϵ_o-net of M. Consequently, $\|x_2 - x_1\| \geq \epsilon_o$ for some $x_2 \in M$. Now consider the set $\{x_1, x_2\}$. Since this set is not an ϵ_o-net of M, there exists $x_3 \in M$ with $\|x_3 - x_i\| \geq \epsilon_o$ for $i = 1, 2$. Continuing the same way, we construct a sequence $\{x_n\}_{n=1}^{\infty}$ such that $\|x_m - x_n\| \geq \epsilon_o$ for $m \neq n$. Thus $\{x_n\}$ does not contain any Cauchy sequence. It follows that no convergent subsequence can be extracted from $\{x_n\}$. This is a contradiction to the compactness of M. Thus, for any $\epsilon > 0$ there exists a finite ϵ-net for M.

Sufficiency. Suppose that for every $\epsilon > 0$ there exists a finite ϵ-net of M and consider a strictly decreasing sequence $\{\epsilon_n\}$, $n = 1, 2, \ldots$, of positive constants such that $\lim_{n \to \infty} \epsilon_n = 0$. Then for each $n = 1, 2, \ldots$ there exists a finite ϵ_n-net of M. If we construct open balls with centers at the points of the ϵ_1-net and radii equal to ϵ_1, then every point of M belongs to one of these balls.

Now let $\{x_n\}_{n=1}^{\infty}$ be a sequence in M. Applying the above argument, there exists a subsequence of $\{x_n\}$, $n = 1, 2, \ldots$, say $\{x_n'\}_{n=1}^{\infty}$, which belongs to one of these ϵ_n-balls. Let $B(y_1)$ be this ball with center y_1. Now we consider the ϵ_2-net of M. The sequence $\{x'\}$ has now a subsequence $\{x_n''\}$, $n = 1, 2, \ldots$, which is contained in some ϵ_2-ball. Let us call this ball $B(y_2)$ (with center at y_2). Coninuing the same way, we obtain a sequence of balls $\{B(y_n)\}_{n=1}^{\infty}$ with centers at y_n, radii ϵ_n and with the following property: the intersection of any finite number of such balls contains a subsequence of $\{x_n\}$. Consequently, we may (and do) choose a subsequence $\{x_{n_k}\}_{k=1}^{\infty}$ of $\{x_n\}$ as follows:

$$x_{n_1} \in B(y_1), \ x_{n_2} \in B(y_2) \cap B(y_1), \ \ldots, \ x_{n_j} \in \bigcap_{i=1}^{j} B(y_i) \text{ with } n_j > n_{j-1} > \ldots > n_1.$$

Since $x_{n_j}, x_{n_k} \in B(y_k)$ for $j \geq k$, we must have

$$\|x_{n_j} - x_{n_k}\| \leq \|x_{n_j} - y_k\| + \|y_k - x_{n_k}\| < 2\epsilon_k$$

Thus, $\{x_{n_j}\}$ is a Cauchy sequence and, since X is complete, it converges to a point in X. This completes the proof.

Proof of Theorem 2.5. Necessity. It suffices to give the proof for $n = 1$. We assume that M is relatively compact. Lemma 2.7 implies now the existence of a finite ϵ-net of M for any $\epsilon > 0$. Let $x_1(t), x_2(t), \ldots, x_n(t)$, $t \in [a, b]$, be the functions of such an ϵ-net. Then for every $f \in M$ there exists

$x_k(t)$ for which $\|f - x_k\|_\infty < \epsilon$. Consequently,

$$|f(t)| \leq |x_k(t)| + |f(t) - x_k(t)| \qquad (2.3)$$

$$\leq \|x_k\|_\infty + \|f - x_k\|_\infty$$

$$< \|x_k\|_\infty + \epsilon$$

Choose now $K = \max \|x_k\|_\infty + \epsilon$. Since each function $x_k(t)$ is uniformly continuous on $[a, b,]$, there exists $\delta_k(\epsilon) > 0$, $k = 1, 2, \ldots$, such that

$$|x_k(t_1) - x_k(t_2)|_\infty < \epsilon \text{ for } |t_1 - t_2| < \delta_k(\epsilon)$$

Let $\delta = \min\{\delta_1, \delta_2, \ldots, \delta_n\}$. Suppose that x is a function in M and let x_j be a function of the ϵ-net for which $\|x - x_j\|_\infty < \epsilon$. Then

$$|x(t_1) - x(t_2)| \leq |x(t_1) - x_j(t_1)| + |x_j(t_1) - x_j(t_2)| + |x_j(t_2) - x(t_2)|$$

$$\leq \|x - x_j\|_\infty + |x_j(t_1) - x_j(t_2)| + \|x - x_j\|_\infty$$

$$< \epsilon + \epsilon + \epsilon = 3\epsilon$$

for all $t_1, t_2 \in [a, b]$ with $|t_1 - t_2| < \delta(\epsilon)$. Consequently, M is equicontinuous. The boundedness of M follows from (2.3).

Sufficiency. Fix $\epsilon > 0$ and pick $\delta = \delta(\epsilon) > 0$ from the condition of equicontinuity. We are going to show the existence of a finite ϵ-net for M. Divide $[a,b]$ into subintervals $[t_{k-1}, t_k]$, $k = 1, 2, \ldots, n$, with $t_0 = a$, $t_n = b$ and $t_k - t_{k-1} < \delta$. Now define a family P of polygons on $[a,b]$ as follows: the function $f:[a, b] \to [-K, K]$ belongs to P if and only if f is a line segment on $[t_{k-1}, t_k]$ for $k = 1, 2, \ldots$, and f is continuous. Thus, if $f \in P$, its vertices (endpoints of its line segments) can appear only at the points $(t_k, f(t_k))$, $k = 0, 1, \ldots, n$. It is easy to see that P is a compact set in $C_1[a,b]$. We show that P is a compact ϵ-net of M. To this end, let $t \in [a,b]$. Then $t \in [t_{j-1}, t_j]$ for some $j = 1, 2, \ldots, n$. If M_j, m_j denote the maximum and the minimum of $x \in M$ in $[t_{j-1}, t_j]$, respectively, then

$$m_j \leq x(t) \leq M_j$$

$$m_j \leq \bar{x}_0(t) \leq M_j$$

where $\bar{x}_0 : [a, b] \to [-K, K]$ is a polygon in P such that $\bar{x}_0(t_k) = x(t_k)$, $k = 1, 2, \ldots, n$. It follows that

$$|x(t) - \bar{x}_0(t)| \leq M_j - m_j < \epsilon$$

Thus P is a compact ϵ-net for M. The reader can now easily check that since P has a finite ϵ-net, say N, the same set N will be a finite 2ϵ-net for M. This completes the proof.

The following two examples give relatively compact subsets of functions in $C_n[a,b]$.

Example 2.8 Let $M \in C_n^1[a, b]$ be such that there exist positive constants K, L with the property:

(i) $\|x(t)\| \leq K$, $t \in [a, b]$
(ii) $\|x'(t)\| \leq L$, $t \in [a, b]$

for every $x \in M$. Then M is a relatively compact subset of $C_n[a,b]$. In fact, the equicontinuity of M follows from the mean value theorem for scalar valued functions or from Exercise 1.10.

Example 2.9 Consider the operator T of Example 1.32. Let $M \subset C_n[a, b]$ be such that there exists $L > 0$ with the property:

$$\|x(t)\| \leq L \text{ for all } x \in M$$

Then the set $S = \{Tu; u \in M\}$ is a relatively compact subset of $C_n[a, b]$. In fact, if

$$N = \sup_{t \in [a,b]} \int_a^b \|K(t, s)\| ds$$

then $\|f\| \leq LN$ for any $f \in S$. Moreover, for $f = Tx$ we have

$$\|f(t_1) - f(t_2)\| = \left\| \int_a^b [K(t_1,s) - K(t_2,s)]x(s)ds \right\|$$

$$\leq L \int_a^b \|K(t_1,s) - K(t_2,s)\| ds$$

This proves the equicontinuity of S.

We now give an example of a bounded sequence in $C_1[0, \infty)$ that is not equicontinuous.

Example 2.10 The sequence $\{f_n(t)\}_{n=1}^\infty$ with

$$f_n(t) = e^{\sin nt}, \quad n = 1, \ldots, \quad t \geq 0$$

does not have any pointwise convergent subsequence on R_+. Thus, it cannot be equicontinuous on R_+.

The next definition is needed in the statement of the Schauder-Tychonov theorem.

Definition 2.11 Let X be a Banach space. Let M be a subset of X. Then M is called "convex" if $\lambda x + (1-\lambda)y \in M$ for any number $\lambda \in [0, 1]$ and any x, $y \in M$.

Theorem 2.12 (Schauder-Tychonov). Let M be a closed, convex subset of a Banach space X. Let $T: M \to M$ be a continuous operator such that TM is a relatively compact subset of X. Then T has a fixed point in M.

It should be noted here that the fixed point of T in the above theorem is not necessarily unique. In the proof of the contraction mapping principle we saw that the unique fixed point of a contraction operator T can be approximated by the terms of a sequence $\{x_n\}_{n=0}^{\infty}$ with $x_j = Tx_{j-1}, j = 1, 2, \ldots$. Unfortunately, no general approximation methods are known for fixed points of operators T as in Theorem 2.12.

Theorem 2.12 suggests the following definition of a compact operator.

Definition 2.13 Let X, Y be two Banach spaces and M a subset of X. An operator $T: M \to Y$ is called "compact" if it is continuous and maps bounded subsets of M into relatively compact subsets of Y.

The example 2.14 below is an application of the Schauder-Tychonov theorem.

Example 2.14 Consider the operator $T: C_n[a, b] \to C_n[a, b]$ defined by

$$(Tx)(t) = f(t) + \int_a^b K(t, s)x(s)ds$$

where $f \in C_n[a, b]$ is fixed and $K:[a, b] \times [a, b] \to M_n$ is continuous. It is easy to show, as in Examples 1.32 and 2.9, that T is continuous on $C_n[a, b]$ and that every bounded set $M \subset C_n[a, b]$ is mapped by T onto the set TM, which is relatively compact. Thus T is compact. Now let

$$M = \{u \in C_n[a, b]; \|u\|_\infty \leq L\}$$

where L is a positive constant. Moreover, let $K + LN \leq L$, where

$$K = \|f\|_\infty, \quad N = \sup_{t \in [a, b]} \int_a^b \|K(t, s)\| ds$$

Then M is a closed, convex, and bounded subset of $C_n[a, b]$ such that $TM \subset M$. By the Schauder-Tychonov theorem, there exists at least one $x_0 \in C_n[a, b]$ such that $x_0 = Tx_0$. For this x_0 we have

$$x_0(t) = f(t) + \int_a^b K(t, s)x_0(s)ds, \quad t \in [a, b]$$

Corollary 2.15 (Brouwer's theorem). Let

$$S = \{u \in R^n;\ \|u\| \leq r\}$$

where r is a positive constant. Let $T\colon S \to S$ be continuous. Then T has a fixed point in S.

Proof. This is a trivial consequence of Theorem 2.12 because every continuous function $f\colon S \to R^n$ is compact.

3. The Leray-Schauder Theorem

Theorem 2.16 (Leray-Schauder). Let X be a Banach space and consider the equation

$$S(x, \mu) - x = 0 \tag{2.4}$$

where:

(i) $S\colon X \times [0, 1] \to X$ is compact in its first variable for each $\mu \in [0, 1]$. Furthermore, if M is a bounded subset of X, then $S(u, \mu)$ is continuous in μ uniformly with respect to $u \in M$; that is, for every $\epsilon > 0$ there exists $\delta(\epsilon) > 0$ with the property: $\|S(u, \mu_1) - S(u, \mu_2)\| < \epsilon$ for every $\mu_1, \mu_2 \in [0, 1]$ with $|\mu_1 - \mu_2| < \delta(\epsilon)$ and every $u \in M$

(ii) $S(x, \mu_0) = 0$ for some $\mu_0 \in [0, 1]$ and every $x \in X$

(iii) if there are any solutions x_μ of the equation (2.4), they belong to some ball of X independently of $\mu \in [0, 1]$.

Then there exists a solution of (2.4) for every $\mu \in [0, 1]$.

The main difficulty in applying the above theorem lies in the verification of the uniform boundedness of the solutions (Condition (iii)). There are no general methods that may be applied to check this condition.

As an application of Theorem 2.16, we provide Example 2.17.

Example 2.17 Let $F\colon R^n \to R^n$ be continuous and such that, for some $r > 0$,

$$\langle F(x), x \rangle \leq \|x\|^2 \text{ whenever } \|x\| > r \tag{2.5}$$

Then $F(x)$ has at least one fixed point in the ball

$$S_r = \{u \in R^n;\ \|u\| \leq r\}$$

Proof. Consider the equation.

$$\mu F(x) - (1 + \epsilon)x = 0 \tag{2.6}$$

with constants $\mu \in [0, 1]$, $\epsilon > 0$. Since every continuous function $F: R^n \to R^n$ is compact, the assumptions of Theorem 2.16 will be satisfied for (2.6), with $S(x, \mu) = [\mu/(1 + \epsilon)]F(x)$, if we show that all possible solutions of (2.6) are in the ball S_r. In fact, let $\bar{x}$ be a solution of (2.6) such that $\|\bar{x}\| > r$. Then we have

$$\langle \mu F(\bar{x}) - (1 + \epsilon)\bar{x}, \bar{x} \rangle = 0$$

or

$$\langle \mu F(\bar{x}), \bar{x} \rangle = (1 + \epsilon) \langle \bar{x}, \bar{x} \rangle = (1 + \epsilon)\|\bar{x}\|^2$$

This implies that

$$\langle F(\bar{x}), \bar{x} \rangle \geq (1 + \epsilon)\|\bar{x}\|^2$$

for some $\bar{x} \in R^n$ with $\|\bar{x}\| > r$, which is a contradiction to (2.5).

It follows, by Theorem 2.16, that for every $\epsilon > 0$ the equation (2.6) has a solution x_ϵ, for $\mu = 1$, such that $\|x_\epsilon\| \leq r$. Let $\epsilon_m = 1/m$, $m = 1, 2, \ldots$, and let $x_{\epsilon_m} = x_m$. Since the sequence $\{x_m\}_{m=1}^\infty$ is bounded, it contains a convergent subsequence $\{x_{m_k}\}_{k=1}^\infty$. Let $x_{m_k} \to x_0$ as $k \to \infty$. Then $x_0 \in S_r$ and

$$F(x_{m_k}) - (1 + (1/m_k))x_{m_k} = 0, \; k = 1, 2, \ldots \tag{2.7}$$

Taking the limit of the left hand side of (2.7) as $k \to \infty$ and using the continuity of F, we obtain

$$F(x_0) = x_0 \tag{2.8}$$

Thus, x_0 is a fixed point of F in S_r.

4. The Inverse Function Theorem

The inverse function theorem is an important tool in the theory of differential equations. It ensures the existence of solutions x of the equation $Tx = y$. Although T is not assumed to be compact and the contraction principle might not be directly applicable, it is shown, in the proof of the inverse function theorem, that the contraction principle can be used indirectly if T has some appropriate differentiability properties.

We start with some definitions concerning the "Frechet differentiability" of the operator T.

Definition 2.18 Let X, Y be Banach spaces and S an open subset of X. Let $f: S \to Y$, $u \in S$ be such that

$$f(u+h) - f(u) = f'(u)h + w(u,h) \qquad (2.9)$$

for every $h \in X$ with $u+h \in S$, where $f'(u): X \to Y$ is a linear operator and

$$\lim_{\|h\| \to 0} \|w(u,h)\|/\|h\| = 0 \qquad (2.10)$$

Then $f'(u)h$ is the "Frechet differential of f at u with increment h", the operator $f'(u)$ is the "Frechet derivative of f at u" (see Lemma 2.21 for the uniqueness of $f'(u)$), and f is called "Frechet diferentiable at $u \in S$."

Defintion 2.19 Assume that X, Y are Banach spaces, S is an open subset of X, and $f: S \to Y$ is continuous and Frechet differentiable at every point of S. Moreover, assume that for every $\epsilon > 0$ there exists $\delta(\epsilon) > 0$ such that

(i) if $u_1, u_2 \in S$ with $\|u_1 - u_2\| \leq \delta(\epsilon)$, then
$$\|[f'(u_1) - f'(u_2)]h\| \leq \epsilon \|h\| \text{ for every } h \in X$$
(ii) $\|w(u,h)\| \leq \epsilon \|h\|$ for every $u \in S$, $h \in X$ with $\|h\| \leq \delta(\epsilon)$.

Then f is called "C-differentiable on S."

In what follows, $S_\mu(x_0)$ is the closed ball with center at x_0 and radius $\mu > 0$.

Definition 2.20 Let S be an open subset of the Banach space X and let f map S into the Banach space Y. Fix a point $u_0 \in S$ and let $f(u_0) = v_0$. Then f is said to be "locally invertible" at (u_0, v_0) if there exist two numbers $\alpha > 0$, $\beta > 0$ with the following property: for every $v \in S_\beta(v_0)$ there exists a unique $u \in S_\alpha(u_0)$ such that $f(u) = v$.

Lemma 2.21 shows the uniqueness of the Frechet derivative.

Lemma 2.21 Let $f: S \to Y$ be given with S an open subset of the Banach space X and Y another Banach space. Suppose further that f is Frechet differentiable at $u \in S$. Then the Frechet derivative of f at u is unique.

Proof. Suppose that $D_1(u)$, $D_2(u)$ are Frechet derivatives of f at u with remainders $w_1(u,h)$, $w_2(u,h)$ respectively. Then we have

$$D_1(u)h + w_1(u,h) = D_2(u)h + w_2(u,h)$$

for every $h \in X$ with $u + h \in S_1$. Here S_1 is some open subset of S containing u. It follows that

$$\|D_1(u)h - D_2(u)h\|/\|h\| = \|w_1(u,h) - w_2(u,h)\|/\|h\| \qquad (2.11)$$

$$\leq \|w_1(u,h)\|/\|h\| + \|w_2(u,h)\|/\|h\|$$

The last member of (2.11) tends to zero as $\|h\| \to 0$. Let

$$Tx = [D_1(u) - D_2(u)]x, \quad x \in X$$

Then T is a linear operator on X such that

$$\lim_{\|x\| \to 0} \|Tx\|/\|x\| = 0$$

Consequently, given $\epsilon > 0$ there exists $\delta(\epsilon) > 0$ such that $\|Tx\|/\|x\| < \epsilon$ for every $x \in X$ with $\|x\| < \delta(\epsilon)$. Given $y \in X$ with $y \neq 0$, let $x = \delta(\epsilon)y/2\|y\|$. Then $\|x\| < \delta(\epsilon)$ and $\|Tx\|/\|x\| < \epsilon$, or $\|Ty\| < \epsilon \|y\|$. Since ϵ is arbitrary, we obtain $Ty = 0$ for every $y \in X$. We conclude that $D_1(u) = D_2(u)$.

The existence of a **bounded** Frechet derivative $f'(u)$ is equivalent to the continuity of f at u. This is the content of the next lemma.

Lemma 2.22 Let $f: S \to Y$ be given, where S is an open subset of a Banach space X and Y another Banach space. Let f be Frechet differentiable at $u \in S$. Then f is continuous at u if and only if $f'(u)$ is a bounded linear operator.

Proof. Let f be continuous at $u \in S$. Then for each $\epsilon > 0$ there exists $\delta(\epsilon) \in (0, 1)$ such that

$$\|f(u+h) - f(u)\| < \epsilon/2, \quad \|f(u+h) - f(u) - f'(u)h\| < (\epsilon/2)\|h\| < \epsilon/2$$

for all $h \in B$ with $u + h \in S$ and $\|h\| < \delta(\epsilon)$.

Therefore,

$$\|f'(u)h\| < (\epsilon/2)\|h\| + \|f(u+h) - f(u)\| < \epsilon$$

for $\|h\| < \delta(\epsilon)$, $u + h \in S$, it follows that the linear operator $f'(u)$ is continuous at the point 0. Exercise 1.4 says that $f'(u)$ is continuous on X. Thus, $f'(u)$ is bounded by Theorem 1.19. The converse is left as an exercise (see Exercise 2.16).

We should note here that the magnitude of the ball S plays no role in the Frechet differentiability of f. This means that, to define the Frechet derivative $f'(u)$ of f at u, we only need to have the equation (2.9) hold for all $u+h$ in a sufficiently small open neighborhood S of the point u.

We quote now a well known theorem of Functional Analysis-the "bounded inverse theorem" (see Schechter [38, Theorem 4.11]).

Theorem 2.23 Let X, Y be Banach spaces and let $T: X \to Y$ be bounded, linear, one-to-one and onto. Then the inverse T^{-1} of T is a bounded linear operator on Y.

We are ready for the inverse function theorem.

Theorem 2.24 (Inverse function theorem). Let X, Y be Banach spaces and S an open subset of X. Let $f: S \to Y$ be C-differentiable on S. Moreover, assume that the Frechet derivative of f is one-to-one and onto at some point $u_0 \in S$. Then the function f is locally invertible at the point $(u_0, f(u_0))$.

Proof. Let $D = f'(u_0)$. Then the operator D^{-1} exists and is defined on Y. Moreover, D^{-1} is bounded (see Theorem 2.23). Thus, the equation $f(u) = v$ is equivalent to the equation $D^{-1}f(u) = D^{-1}v$. Fix $v \in Y$ and define the operator U on S as follows:

$$Uu = u + D^{-1}[u - f(u)], \quad u \in S \tag{2.11}$$

Obviously, the fixed points of the operator U are solutions to the equation $f(u) = v$. We first determine a closed ball inside S with center at u_0 on which U is a contraction operator. To this end, fix $\epsilon \in (0, 1/4\|D^{-1}\|)$ and let $\delta(\epsilon) > 0$ be such that

$$\|[f'(u_0) - f'(u_2)](u_1 - u_2)\| \leq \epsilon \|u_1 - u_2\| \tag{2.12}$$

$$\|w(u_2, u_1 - u_2)\| \leq \epsilon \|u_1 - u_2\| \tag{2.13}$$

for every $u_1, u_2 \in S$ with $\|u_1 - u_0\| \leq \delta(\epsilon)/2$, $\|u_2 - u_0\| \leq \delta(\epsilon)/2$.

This is possible by virtue of the C-differentiability of the function f. Thus, we have

$$\|Uu_1 - Uu_2\| = \|u_1 - u_2 - D^{-1}[f(u_1) - f(u_2)]\|$$

$$= \|D^{-1}f'(u_0)(u_1 - u_2) - D^{-1}f'(u_2)(u_1 - u_2)$$

$$- D^{-1}w(u_2, u_1 - u_2)\| \tag{2.14}$$

$$\leq \|D^{-1}\| \|[f'(u_0)-f'(u_2)](u_1-u_2)\| + \|D^{-1}\| \|w(u_2, u_1-u_2)\|$$

$$\leq \epsilon \|D^{-1}\| \|u_1-u_2\| + \epsilon \|D^{-1}\| \|u_1-u_2\|$$

$$\leq (1/2)\|u_1-u_2\|$$

for every $u_1, u_2 \in S$ as above. It follows that U is a contraction operator on the ball $S_\alpha(u_0)$, where $\alpha = \delta(\epsilon)/2$, $\epsilon < 1/(4\|D^{-1}\|)$.

Now we determine a constant $\beta > 0$ such that $US_\alpha(u_0) \subset S_\alpha(u_0)$ whenever $v \in S_\beta(v_0)$. Here $v_0 = f(u_0)$. In fact, we have

$$\|Uu_0 - u_0\| \leq \|D^{-1}\| \|v - f(u_0)\| = \|D^{-1}\| \|v - v_0\|$$

$$\leq \delta(\epsilon)/4$$

whenever

$$\|v - v_0\| \leq \delta(\epsilon)/(4\|D^{-1}\|) = \beta$$

Furthermore,

$$\|Uu - u_0\| = \|Uu - Uu_0\| + \|Uu_0 - u_0\|$$

$$\leq (1/2)\|u - u_0\| + \delta(\epsilon)/4 \leq \delta(\epsilon)/4 + \delta(\epsilon)/4 = \delta(\epsilon)/2$$

for any $u \in S_\alpha(u_0)$. We have shown that f is locally invertible at $(u_0, f(u_0))$, and that for any $v \in S_\beta(v_0)$ there exists a unique $u \in S_\alpha(u_0)$ with $f(u) = v$.

Remark 2.25 It is important to note that Condition (ii) in Definition 2.19 holds in a sufficiently small neighborhood of any point in S if Condition (ii) is satisfied on S. The proof of this fact involves some elementary properties of Riemann integrals of continuous functions with values in Banach spaces, and it is therefore omitted. However, in order to make the conditions of the inverse function theorem easier to check, we state the following version of it.

Theorem 2.26 Let the assumptions of Theorem 2.24 be satisfied with the C-differentiability of the function f replaced by Condition (i) of Definition 2.19. Then the conclusion of Theorem 2.24 remains valid.

Example 2.27 Let $J = [a, b]$ and let

$$S_r = \{u \in R^n; \|u\| < r\}$$

$$S' = \{x \in C_n(J); \|x\|_\infty < r\}$$

where r is a positive number.

We consider a continuous function $F: J \times \bar{S}_r \to R^n$ and the operator $U: S' \to C_n(J)$ defined as follows:

$$(Ux)(t) = F(t, x(t)), \ t \in J, \ x \in S'$$

We first note that U is continuous on S'. In fact, since F is uniformly continuous on the compact set $J \times \bar{S}_r$, for every $\epsilon > 0$ there exists $\delta(\epsilon) > 0$ such that

$$\|F(t, u) - F(t, v)\| < \epsilon$$

for every $u, v \in \bar{S}_r$ with $\|u - v\| < \delta(\epsilon)$ and every $t \in J$. This implies that

$$\|Ux - Uy\|_\infty < \epsilon$$

whenever $x, y \in S'$ with $\|x - y\|_\infty < \delta(\epsilon)$. In order to compute the Frechet derivative of U, we assume that the Jacobian matrix

$$F_x(t, u) = [(\partial F_i / \partial u_j)(t, u)], \ i = 1, 2, \ldots, n$$

exists and is continuous on $J \times S_r$. Then, given two functions $x_0 \in S'^1 (0 < r_1 < r)$, $h \in C_n(J)$ such that $x_0 + h \in S'^1$, we have

$$\sup_{t \in J} \|F(t, x_0(t) + h(t)) - F(t, x_0(t)) - F_x(t, x_0(t))h(t)\|$$

$$\leq \sup_{t \in J} \{\|[(\partial F_i/\partial x_j)(t, x_0(t) + \theta_i h(t))] - F_x(t, x_0(t))]\|\} \|h\|_\infty \qquad (2.15)$$

where θ_i, $i = 1, 2, \ldots, n$, are functions of t lying in the interval $(0, 1)$. In (2.15) we have used the mean value theorem for real valued functions on S_r as follows:
$$F_i(t, x_0(t) + h(t)) - F_i(t, x_0(t)) = <\nabla F_i(t, z_i(t)), h(t)>, \ i = , 2, \ldots, n$$

where $z_i(t) = x_0(t) + \theta_i h(t)$. Form the uniform continuity of $\nabla F_i(t, u)$, $i = i, 2, \ldots, n$, on $J \times S_{r_1}$, it follows that the Frechet derivative $U'(x_0)$ exists and is a bounded linear operator given by the formula

$$[U'(x_0)h](t) = F_x(t, x_0(t))h(t) \qquad (2.16)$$

for every $h \in C_n(J)$ and every $t \in J$. Since for every $x \in S_r$ there exists some S_{r_1} such that x, $x + h \in S_{r_1}$ for sufficiently small $\|h\|$, (2.16) holds for all $x_0 \in S^r$. For the same reason, it suffices to have F defined just on $J \times S_r$.

EXERCISES

2.1. Let $f_n: [a, b] \to R$, $n = 1, 2, \ldots$, be a sequence of continuous functions converging to zero as $n \to \infty$ uniformly on $[a, b]$. Show that the functions form a relatively compact set in $C_1[a, b]$.

2.2. Consider the operator

$$(Tx)(t) = y(t) \int_0^\infty e^{3s} x(s) ds, \quad t \in R_+$$

where $y \in C_1(R_+)$ satisfies

$$\sup_{t \in R_+} \{e^{4t} |y(t)|\} < 1$$

Obviously, T is not well defined for all $x \in C_1(R_+)$. We consider the "weighted" norm

$$\|x\|_e = \sup_{t \in R_+} \{e^{4t} |x(t)|\} \tag{2.15}$$

and the space

$$C_e = \{f \in C_1(R_+); \ \|f\|_e < +\infty\}$$

Show that C_e is a Banach space with the norm (2.15) and that T maps every closed ball of C_e with center at zero into itself. Using the contraction principle, show that $x(t) \equiv 0$, $t \in R_+$, is the only solution to the equation

$$x(t) = y(t) \int_0^\infty e^{3s} x(s) ds, \quad t \in R_+$$

2.3. Let X, Y, Z be Banach spaces. Let $T: X \to Y$, $U: Y \to Z$ be compact. Show that the operator $UT: X \to Z$ is compact. Here UT is the composition of U and T.

2.4. Let X be a Banach space and $S_0: X \to X$ a compact operator. Let

$$\|S_0 x\| \leq \lambda \|x\| + m$$

where $\lambda < 1$, m are positive constants. Show that S_0 has at least one fixed point in X. [Hint. Apply the Leray-Schauder theorem to the operator $S(x, \mu) = \mu S_0 x$, $\mu \in [0, 1]$. Actually, here we may also use the Schauder-Tychonov theorem. (How?)]

2.5. Let $M \subset C_n^i$ have the following properties:

(i) For every $\epsilon > 0$ there exists $\delta(\epsilon) > 0$ such that

$$\|u(t) - u(t')\| < \epsilon$$

for every $t, t' \in R_+$ with $|t - t'| < \delta(\epsilon)$ and every $u \in M$

(ii) There exists a constant $L > 0$ such that $\|u\|_\infty \leq L$ for every $u \in M$

(iii) Let l_u denote the limit of $u(t)$ as $t \to +\infty$. Then for every $\epsilon > 0$ there exists $Q(\epsilon) > 0$ such that

$$\|u(t) - l_u\| < \epsilon \text{ for every } t > Q(\epsilon), u \in M.$$

Show that M is relatively compact.

2.6. Let X, Y be Banach spaces and S an open subset of X. Let $T: X \to Y$ be a bounded linear operator and $f: S \to Y$ have Frechet derivative $f'(u_0)$ at $u_0 \in S$. Show that $Tf: X \to Y$ is Frechet differentiable at u_0 with Frechet derivative

$$(Tf)'(u_0) = T(f'(u_0))$$

2.7. Consider the operator

$$(Tx)(t) = \int_0^t [1 - x^2(s)]^{1/3} ds$$

defined on $C_1[0, 1]$. Show that T has a fixed point.

2.8. Let the operator T be defined on $C_1(R_+)$ as follows:

$$(Tx)(t) = f(t) + (1/2) \int_0^t e^{-s} \sin(x(s)) ds$$

Here $f \in C_1(R_+)$ is a given function. Show that T has a unique fixed point $x_0 \in C_1(R_+)$.

2.9. Let $F: R^n \to R^n$ be continuous and such that

$$\|F(x)\| \leq \|x\| \text{ whenever } \|x\| > r$$

where r is a positive constant. Show that F has a fixed point in the ball $\{u \in R^n;\ \|u\| \leq r\}$.

2.10. Let $T: C_3[a, b] \to C_1[a, b]$ be defined as follows:

$$(Tx)(t) = x_1^2(t) + \sin(x_2(t)) - x_1(t)\exp(x_3(t)),\ t \in [a, b]$$

where $x(t) = (x_1(t), x_2(t), x_3(t))$. Find a formula for the Frechet derivative $T'(x_0)$ at any $x_0 \in C_3[0, 1]$.

2.11. Let

$$F(u) = \begin{bmatrix} u_1^2 \\ u_1 + u_2 \\ \sin u_3 \end{bmatrix}$$

$u = (u_1, u_2, u_3) \in R^3$. Show that F is Frechet differentiable at every $x \in R^n$ and find its Frechet derivatives.

2.12. Let S be an open subset of a Banach space X with norm $\|\cdot\|$. Let $f: S \to X$ have Frechet derivative $f'(u_0)$ at some point $u_0 \in S$. Show that if $\|\cdot\|_a$ is another norm of X, equivalent to $\|\cdot\|$, then the Frechet derivative of f w.r.t. $\|\cdot\|_a$ is $f'(u_0)$.

2.13. Let the operator T be defined as follows:

$$(Tx)(t) = f(t) + \int_a^t K(s)F(x(s))ds$$

where $f \in C_1[a, b]$, $K \in C_1[a, b]$ are known functions and $F: R \to R$ is continuous. Provide further conditions on F so that the Frechet derivative of T at $x_0 \in C_1[a, b]$ exists, and provide a formula for it.

2.14. In the proof of the inverse function theorem we found two closed balls $S_\alpha(u_0)$, $S_\beta(v_0)$ such that for every $v \in S_\beta(v_0)$ there is a unique $u = u(v) \in S_\alpha(u_0)$ such that $f(u) = v$. Show that the function $v \to u(v)$ is continuous on $S_{\beta_1}(v_0)$, where β_1 is some number in $(0, \beta]$.

2.15. Consider the function

$$F(u) = \begin{bmatrix} u_1 + u_2^2 \\ u_3 \\ u_1 - u_3 \end{bmatrix}$$

$u = (u_1, u_2, u_3) \in R^3$. Show that there is a closed ball $S \in R^3$ with center at $(0, 1, 0)$ and a closed ball $S \in R^3$ with center at $(1, 0, 0)$ with the property: for every $y \in S_1$ there exists a unique $x \in S$ such that $F(x) = y$.

2.16. In the setting of Lemma 2.22, prove that if $f'(u)$ is bounded, then f is continuous at u.

CHAPTER THREE

EXISTENCE, UNIQUENESS, AND CONTINUATION FOR DIFFERENTIAL SYSTEMS;
BASIC THEORY OF LINEAR SYSTEMS

In this chapter we study systems of the form

$$x' = F(t, x) \qquad (3.1)$$

where $F: J \times M \to R^n$ is continuous. Here J is an interval of R and M is a subset of R^n.

In Section 1 we state and prove the fundamental theorem of Peano. This theorem ensures the existence of local solutions of (3.1) under the mere assumption of continuity of F. The uniqueness of the local solution follows from an assumed Lipschitz condition on F. This is the Picard-Lindelof theorem (Theorem 3.2), which we prove by using the method of successive approximations as in the scalar case.

In Section 2 we concern ourselves with the problem of continuation of solutions of (3.1). Roughly speaking, we show that the boundedness of the solution $x(t)$, $t \in [a, b)$, or the boundedness of the function $F(t, u)$, implies the continuation of $x(t)$ to point b. A similar situation exists for left end-points of existence of $x(t)$.

Section 3 is devoted to the establishment of some elementary properties of linear systems. These properties are used in later chapters in order to obtain further information about such systems or perturbed linear systems.

1. Existence, Uniqueness

Theorem 3.1 (Peano) Let (t_0, x_0) be a given point in $R \times R^n$. Let $J = [t_0 - a, t_0 + a]$, $D = \{x \in R^n; \|x - x_0\| \leq b\}$, where a, b are two positive numbers, and for the system (3.1) assume the following: $F: J \times D \to R^n$ is continuous with $\|F(t, u)\| \leq L$, $(t, u) \in J \times D$, where L is a positive constant. Then there exists a solution $x(t)$ of (3.1) with the following property: $x(t)$ is defined and satisfies (3.1) on $S = \{t \in J; |t - t_0| \leq \alpha\}$ with $\alpha = \min\{a, b/L\}$. Moreover, $x(t_0) = x_0$ and $\|x(t) - x_0\| \leq b$ for all $t \in [t_0 - \alpha, t_0 + \alpha]$.

Proof. To prove this result, we apply the Schauder-Tychonov theorem (Theorem 2.12). To this end, we consider first the operator

$$(Tu)(t) = x_0 + \int_{t_0}^{t} F(s, u(s)) ds, \quad t \in J_1$$

where $J_1 = [t_0, t_0 + \alpha]$.

Let

$$D_0 = \{u \in C_n(J_1); \|u - x_0\|_\infty \leq b\}$$

We first show that D_0 is convex. In fact, let $u_1, u_2 \in D_0$ and $\lambda \in [0, 1]$. Then we have, for $t \in J_1$,

$$\|\lambda u_1(t) + (1 - \lambda) u_2(t) - x_0\| = \|(\lambda u_1(t) - \lambda x_0) + (1 - \lambda) u_2(t) - (1 - \lambda) x_0\|$$

$$\leq \lambda \|u_1(t) - x_0\| + (1 - \lambda) \|u_2(t) - x_0\| \leq b$$

Obviously, D_0 is closed. To show that T maps D_0 into itself, let $u \in D_0$. Then

$$\|(Tu)(t) - x_0\| \leq \int_{t_0}^{t_0 + \alpha} \|F(s, u(s))\| ds \leq \alpha L \leq b$$

Thus, $TD_0 \subset D_0$. To show that TD_0 is equicontinuous, let $u \in D_0$. Then

$$\|(Tu)(t_1) - (Tu)(t_2)\| \leq \left| \int_{t_1}^{t_2} \|F(s, u(s))\| ds \right| \leq L |t_1 - t_2|$$

for any $t_1, t_2 \in J_1$. According to Theorem 2.5, TD_0 is relatively compact.

Given $\epsilon > 0$ there exists $\delta(\epsilon) > 0$ such that $u, v \in D_0$, $\|u - v\|_\infty < \delta(\epsilon)$ imply that $\|F(\cdot, u(\cdot)) - F(\cdot, v(\cdot))\|_\infty < \epsilon$. This follows from the uniform continuity of F on the set $D_1 = [t_0, t_0 + \alpha] \times \{x \in R^n; \|x - x_0\| \leq b\}$. For the proof of the continuity of T on D_0, let $u_n, u \in D_0$ be such that $\|u_n - u\|_\infty \to 0$. Then, given $\epsilon > 0$, there exists $N(\epsilon) > 0$ with $\|u_n - u\|_\infty < \delta(\epsilon)$ for $n > N(\epsilon)$. Thus, we easily obtain $\|Tu_n - Tu\|_\infty < \alpha \epsilon$ for $n > N(\epsilon)$, and this proves the continuity of T on

D_0. The Schauder-Tychonov theorem applies now and ensures the existence of a fixed point of T; that is, a function $u \in D_0$ such that

$$u(t) = x_0 + \int_{t_0}^{t} F(s, u(s))ds \tag{3.2}$$

for every $t \in [t_0, t_0 + \alpha]$.

The same method can be applied to show the existence of a solution $\bar{u}(t)$ of (3.2) on the interval $[t_0 - \alpha, t_0]$. Both functions $u(t)$, $\bar{u}(t)$ satisfy (3.1) on their respective domains. Now consider the function

$$x(t) = \begin{cases} \bar{u}(t), & t \in [t_0 - \alpha, t_0] \\ u(t), & t \in [t_0, t_0 + \alpha] \end{cases} \tag{3.3}$$

This function satisfies the equation

$$x(t) = x_0 + \int_{t_0}^{t} F(s, x(s))ds, \quad t \in J \tag{3.4}$$

and it is the desired solution to (3.1).

The uniqueness of the above solution can be achieved by assuming a Lipschitz condition on F w.r.t. its second variable. This is the content of Theorem 3.2.

Theorem 3.2 (Picard-Lindelof) Consider System (3.1) under the assumptions of Theorem 3.1. Let $F: J \times D \to R^n$ satisfy

$$\|F(t, x_1) - F(t, x_2)\| \le k \|x_1 - x_2\|$$

for every $(t, x_1), (t, x_2) \in J \times D$, where k is a positive constant. Then there exists a unique solution $x(t)$ satisfying the conclusion of Theorem 3.1.

Proof. We give a proof of the existence of $x(t)$ independent of the result in Theorem 3.1. The method employed here uses successive approximations as in the scalar case. In fact, consider the sequence

$$y_0(t) = x_0$$
$$y_{m+1}(t) = x_0 + \int_{t_0}^{t} F(s, y_m(s))ds, \quad m = 1, 2, \ldots$$

for $t \in [t_0, t_0 + \alpha]$. Then it is easy to show that, for $m = 1, 2, \ldots$,

$$\|y_m(t) - x_0\| \le b$$
$$\|y_{m+1}(t) - y_m(t)\| \le Lk^m(t - t_0)^{m+1}/(m+1)!$$

for every $t\in[t_0, t_0 + \alpha]$. It follows that the series

$$y_0 + \sum_{m=0}^{\infty} [y_{m+1}(t) - y_m(t)] \tag{3.5}$$

converges uniformly on $[t_0, t_0 + \alpha]$ to a function $y(t)$, $t\in[t_0, t_0 + \alpha]$. In fact, this is a consequence of Weierstrass' M-Test and the fact that the terms of the above series are bounded above by the corresponding terms of the series

$$\|y_0\| + \frac{L}{k} \sum_{m=0}^{\infty} \frac{(k\alpha)^{m+1}}{(m+1)!}$$

which converges to the number $\|y_0\| + \frac{L}{k}(e^{k\alpha} - 1)$. Since the partial sum S_n of the series in (3.5) equals $y_n(t)$, $n = 0, 1, 2, \ldots$, we obtain that the sequence $\{y_n(t)\}$ converges as $n\to\infty$ to a function $y(t)$ uniformly on $[t_0, t_0 + \alpha]$. This function $y(t)$ satisfies

$$y(t) = x_0 + \int_{t_0}^{t} F(s, y(s))ds, \quad t\in[t_0, t_0 + \alpha]$$

Thus, $y(t_0) = x_0$, $\|y(t) - x_0\| \leq b$, $t\in[t_0, t_0 + \alpha]$, and $y(t)$ satisfies (3.1) on the above interval. This process can now be repeated on the interval $[t_0 - \alpha, t_0]$ to obtain a solution $\bar{y}(t)$ with similar properties on this interval. The function $x(t)$ which is identical to $y(t)$ on $[t_0, t_0 + \alpha]$ and to $\bar{y}(t)$ on $[t_0 - \alpha, t_0]$ is a solution on $[t_0 - \alpha]$. Now let $x_1(t)$ be another solution having the same properties as $x(t)$. Then we can use the equation

$$x_1(t) = x_0 + \int_{t_0}^{t} F(s, x_1(s))ds$$

to obtain, by induction, that

$$\|y_n(t) - x_1(t)\| \leq \frac{L}{k} \frac{(k\alpha)^{n+1}}{(n+1)!}, \quad t\in[t_0, t_0 + \alpha]$$

Taking limits as $n\to\infty$, we obtain $x(t) = x_1(t)$, $t\in(t_0, t_0 + \alpha]$. Similarly one argues on the interval $[t_0 - \alpha, t_0]$. This completes the proof of the theorem.

It is obvious that the interval J in Theorems 3.1 and 3.2 may be replaced by one of the intervals $[t_0 - a, t_0]$, $[t_0, t_0 + a]$ in which case the solution found to exist will be defined on $[t_0 - \alpha, t_0]$ or $[t_0, t_0 + \alpha]$, respectively.

It is quite important to mention here that a certain weaker type of solution of the equation (3.1) can exist although F has some discontinuities w.r.t. its first variable t. Roughly speaking, if these discontinuities are integrable, then a unique solution exists as in the above theorem satisfying the integral equation

$$x(t) = x_0 + \int_a^t F(s, x(s))ds \tag{3.6}$$

with an improper integral.

This case is partially covered by the theorem below, which holds for functions F that may have a discontinuity at $t = a$.

Theorem 3.3 Let $F: (a, b] \times D \to R^n$ be continuous, where $D = \{u \in R^n; \|u - x_0\| \le r\}$. Here x_0 is a fixed point in R^n and r is a positive constant. Let

$$\|F(t, u_1) - F(t, u_2)\| \le l(t)\|u_1 - u_2\|$$

$$\|F(t, u)\| \le m(t)$$

for every $t \in (a, b]$, $u, u_1, u_2 \in D$, where $l, m: (a, b] \to R_+$ are continuous and integrable (in the improper sense) on $(a, b]$. Choose the number b_1 $(a < b_1 < b)$ so that

$$L = \int_{a^+}^{b_1} l(t)dt < 1, \quad \int_{a^+}^{b_1} m(t)dt \le r$$

Then the integral equation (3.6) has a unique solution $x(t)$ on the interval $[a, b_1]$. This solution satisfies $x(a) = x(a^+) = x_0$ and the system (3.1) on the interval $(a, b_1]$.

Proof. The proof follows the steps of the previous theorem by considering the successive approximations $y_n(t)$, $n = 0, 1, \ldots$. We only show the uniqueness of the solution $x(t)$. In fact, let $y(t)$ be another solution of the integral equation (3.6) on $[a, b_1]$. Then we have

$$\|x(t) - y(t)\| \le \int_{a^+}^t l(s)\|x(s) - y(s)\|ds, \quad t \in [a, b_1) \tag{3.7}$$

which implies

$$\|x - y\|_\infty \le \int_{a^+}^{b_1} l(s)ds \|x - y\|_\infty = L\|x - y\|_\infty$$

Since $L < 1$, we have $\|x - y\|_\infty = 0$, or $x(t) \equiv y(t)$, $t \in [a, b_1]$.

The uniqueness part of Theorem 3.2 can be shown with the help of an elementary inequality-"Gronwall's inequality." In Theorem 3.2 we gave a proof using successive approximations in order to exhibit the method in R^n. However, Gronwall's inequality will be needed several times in subsequent chapters. We state the relevant lemma here for convenience.

Lemma 3.4 (Gronwall's inequality) Let $u, g: [a, b] \to R_+$ be continuous and such that

$$u(t) \le K + \int_a^t g(s)u(s)ds, \quad t \in [a, b]$$

$$u(t) \leq K\exp\{\int_a^t g(s)ds\}, \ t\in[a, b]$$

The proof is left as an exercise (see Exercise 3.1).

2. Continuation

In this section we study the problem of continuation (or extendability) of the solutions whose local existence is ensured by Theorems 3.1 and 3.2. In what follows, a domain is an open, connected set. We have the following definition.

Definition 3.5 A solution $x(t)$, $t\in[a, b)$, $a<b<+\infty$, of System (3.1) is said to be "continuable to $t = b$", if there exists another solution $\bar{x}(t)$, $t\in[a, c]$, $c\geq b$, of the system (3.1) such that $\bar{x}(t) = x(t)$, $t\in[a, b)$. A solution $x(t)$, $t\in[a, b)$ $(a<b<+\infty)$, of the system (3.1) is said to be "continuable to $t = c$" $(b < c < +\infty)$, if it is continuable to $t = b$ and whenever we assume that $x(t)$ is a solution on $[a, d)$, for any $d\in(b, c]$, we can show that $x(t)$ is continuable to the point $t = d$. Such a solution is "continuable to $+\infty$" if it is continuable to $t = c$ for any $c > b$. Similarly one defines continuation to the left.

In what follows, we also use the word "extendable" instead of "continuable."

Theorem 3.6 Suppose that D is a domain of $R\times R^n$ and that $F: D\to R^n$ is continuous. Let (t_0, x_0) be a point in D and assume that the system (3.1) has a solution $x(t)$ defined on a finite interval (a, b) with $t_0\in(a, b)$ and $x(t_0) = x_0$. Then if F is bounded on D, the limits

$$x(a^+) = \lim_{t\to a^+} x(t),$$

$$x(b^-) = \lim_{t\to b^-} x(t) \qquad (3.8)$$

exist as finite vectors. If the point $(a, x(a^+))$ $((b, x(b^-)))$ is in D, then $x(t)$ is continuable to the point $t = a$ $(t = b)$.

Proof. In order to show that the first limit in (3.8) exists, we first note that

$$x(t) = x_0 + \int_{t_0}^t F(s, x(s))ds, \ t\in(a, b) \qquad (3.9)$$

Now let $\|F(t, x)\| \leq L$ for $(t, x)\in D$, where L is a positive constant. Then if t_1, $t_2\in(a, b)$, we obtain

$$\|x(t_1) - x(t_2)\| \leq \left|\int_{t_1}^{t_2} \|F(s, x(s))\|\ ds\right| \leq L|t_1 - t_2|$$

Thus, $x(t_1) - x(t_2)$ converges to zero as t_1, t_2 converge to the point $t = a$ from the right. Applying the Cauchy criterion for functions, we obtain our assertion. Similarly one argues for the second limit of (3.8).

Let us assume now that the point $(b, x(b^-))$ belongs to D and consider the function.
$$\bar{x}(t) = \begin{cases} x(t), & t \in (a, b) \\ x(b^-), & t = b \end{cases}$$

This function is a solution of (3.1) on $(a, b]$. In fact, (3.9) implies
$$\bar{x}(t) = x_0 + \int_{t_0}^{t} F(s, \bar{x}(s)) ds, \quad t \in (a, b]$$
which in turn implies the existence of the left-hand derivative $\bar{x}'_-(b)$ of $\bar{x}(t)$ at $t = b$.

Thus we have
$$\bar{x}'_-(b) = F(b, \bar{x}(b))$$
which completes the proof for $t = b$. A similar argument holds for $t = a$.

It should be noted that if the point $(a, x(a^+))$ is not in D, but $F(a, x(a^+))$ can be defined so that F is continuous at $(a, x(a^+))$, then $x(t)$ is continuable to $(a, x(a^+))$. A similar situation exists at $(b, x(b^-))$.

In what follows, we are mainly concerned with the continuation to the right of the interval of existence of the solution. The reader should bear in mind that corresponding results cover the continuation to the left of this interval. The following continuation theorem is needed for the proof of of Theorem 3.8. More general theorems can be found in Chapter Five.

Theorem 3.7 Let $F: [a, b] \times R^n \to R^n$ be continuous and such that $\|F(t, x)\| \leq L$ for all $(t, x) \in [a, b] \times R^n$, where L is a positive constant. Then every solution $x(t)$ of (3.1) is continuable to the point $t = b$.

Proof. Let $x(t)$ be a solution of (3.1) passing through the point $(t_0, x_0) \in [a, b] \times R^n$. Assume that $x(t)$ is defined on the interval $[t_0, c)$ where c is some point with $c \leq b$. Then, as in the proof of Theorem 3.3, $x(c^-)$ exists and $x(t)$ is continuable to the point $t = c$. If $c = b$ the proof is complete. If $c < b$, then Peano's theorem (Theorem 3.1), applied on $[c, b] \times D$, with D a sufficiently large closed ball with center at $x(c^-)$, ensures the existence of a solution $\bar{x}(t)$, $t \in [c, b]$, such that $\bar{x}(c) = x(c^-)$. Thus, the function
$$x_0(t) = \begin{cases} x(t), & t \in [t_0, c] \\ \bar{x}(t), & t \in [c, b] \end{cases}$$

is the desired continuation of $x(t)$.

Theorem 3.8 says that the boundeness of every solution through a certain point, in a certain sense, implies its extendability.

Theorem 3.8 Let $F: [a, b] \times M \to R^n$ be continuous, where M is the closed ball $S_r = \{u \in R^n; \|u\| \leq r\}, r > 0$ (or R^n). Assume that $(t_0, x_0) \in [a, b] \times M$ is given and that every solution $\bar{x}(t)$ of (3.1) passing through (t_0, x_0) satisfies $\|x(t)\| < \lambda$ as long as it exists to the right of t_0. Here $0 < \lambda \leq r$ (or $0 < \lambda < \infty$). Then every solution $x(t)$ of (3.1) with $x(t_0) = x_0$ is continuable to the point $t = b$.

Proof. We give the proof for $M = S_r$. If $M = R^n$ the same proof holds and it is even easier. Let $x(t)$ be a solution of (3.1) with $x(t_0) = x_0$ and assume that $x(t)$ is defined on $[t_0, c)$ with $c < b$. Since F is continuous on $[a, b] \times S_\lambda$, where

$$S_\lambda = \{u \in R^n; \|u\| \leq \lambda\}$$

there exists $L > 0$ such that $\|F(t, x)\| \leq L$ for all $(t, x) \in [a, b] \times S_\lambda$. Now consider the function

$$F_1(t, x) = \begin{cases} F(t, x), & (t, x) \in [a, b] \times S_\lambda \\ \dfrac{\lambda}{\|x\|} F\left(t, \dfrac{\lambda x}{\|x\|}\right), & t \in [a, b], \|x\| \geq \lambda \end{cases}$$

It is easy to see that F_1 is continuous and such that $\|F_1(t, x)\| \leq L$ on $[a, b] \times R^n$. Consequently, Theorem 3.7 implies that every solution of the system

$$x' = F_1(t, x) \qquad (3.10)$$

is continuable to $t = b$. Naturally, $x(t)$ is a solution of (3.10) defined on $[t_0, c)$ because $F_1 = F$ for $\|x\| \leq \lambda$. Therefore, there exists a solution $x_1(t), t \in [t_0, b]$, of (3.10) such that $x_1(t) = x(t), t \in [t_0, c)$. Assume that there exists $t_1 \in [c, b]$ such that $\|x_1(t_1)\| = \lambda$. Then, for some $t_2 \in [c, t_1], \|x_1(t_2)\| = \lambda$ and $\|x_1(t)\| < \lambda$ for all $t \in [t_0, t_2)$. Obviously, $x_1(t)$ satisfies the system (3.1) on $[t_0, t_2]$. This is a contradicion to our assumption. Thus, $\|x_1(t)\| < \lambda$ for all $t \in [t_0, b]$, which implies that $x(t)$ is continuable to the point $t = b$.

The following theorem is an important tool in dealing with various problems on infinite intervals. It guarantees the existence of a bounded solution on an infinite interval under the assumption of the existence of a local solution, on any finite subinterval, which is bounded by a fixed positive constant.

Theorem 3.9 Let the function $F: R \times R^n \to R$ ($F: R_+ \times R^n \to R^n$) be continuous. For each $m = 1, 2, \ldots$, assume the existence of a solution $x_m(t)$, $t \in [-m, m]$ ($t \in [0, m]$) of (3.1) such that $\|x_m\|_\infty \leq K$, where K is a fixed positive constant. Then the system (3.1) has at least one solution $x(t)$, $t \in R$ ($t \in R_+$), such that $\|x\|_\infty \leq K$.

Proof. We give the proof for $t \in R$. The proof on R_+ follows similarly. Let

$$q_m = \sup_{\substack{|t| \leq m \\ \|u\| \leq K}} \|F(t, u)\|, \quad m = 1, 2, \ldots$$

Then for every $m = 1, 2, \ldots$ we have

$$x_m(t) = x_m(0) + \int_0^t F(s, x_m(s))ds, \quad t \in [-m, m]$$

This yields

$$\|x_m(t) - x_m(t')\| \leq q_m |t - t'|, \quad t, t' \in [-m, m]$$

It follows that the sequence $\{x_m(t)\}_{m=k}^\infty$ is uniformly bounded and equicontinuous on the interval $[-k, k]$ for any $k = 1, 2, \ldots$. Theorem 2.5 implies the existence of a subsequence $\{x_{m1}(t)\}_{m=1}^\infty$ of $\{x_m(t)\}$ which converges uniformly to a function $\bar{x}_1 \in C_n[-1, 1]$ as $m \to \infty$. This sequence $\{x_{m1}(t)\}$ has a sequence $\{x_{m2}(t)\}_{m=1}^\infty$ which converges uniformly to a function $\bar{x}_2 \in C_n[-2, 2]$. By induction, we can construct a subsequence $\{x_{m, j+1}(t)\}_{m=1}^\infty$ of the sequence $\{x_{mj}(t)\}_{m=1}^\infty$ converging uniformly to the function $\bar{x}_j \in C[-j, j]$. All functions $\bar{x}_m$ satisfy $\|\bar{x}_m\|_\infty \leq K$, $m = 1, 2, \ldots$. The diagonal sequence $\{x_{mm}(t)\}$ is defined for all large m on the interval $[-r, r]$ for any $r > 0$) and converges uniformly to a function $x(t)$ there. We show that $x(t)$ is the desired solution. In fact, given $r > 0$, take $m_0 > r$. Then $\{x_{mm}(t)\}$ is defined on the interval $[-r, r]$ for all $m \geq m_0$, and we have

$$x_{mm}(t) = x_{mm}(0) + \int_0^t F(s, x_{mm}(s))ds \qquad (3.11)$$

for $t \in [-r, r]$. Taking the limit of each side of (3.11) as $m \to \infty$, we obtain

$$x(t) = x(0) + \int_0^t F(s, x(s))ds$$

or

$$x'(t) = F(t, x(t)), \quad t \in [-r, r]$$

Since r is arbitrary, the proof is complete.

3. Linear Systems

In this section we consider systems of the form

$$x' = A(t)x \qquad (S)$$

$$x' = A(t)x + f(t) \qquad (S_f)$$

Here $A(t)$ is an $n \times n$ matrix of continuous functions on a real interval J and $f(t)$ is an n-vector of continuous functions on J. For the system (S_f) we have the following general existence theorem. An independent proof is given in order to exhibit the method.

Theorem 3.10 Let $A: J \to M_n$, $f: J \to R^n$ be continuous. Let (t_0, x_0) be a point in $J \times R^n$. Then there exists a unique solution $x(t)$ of (S_f) which is defined on J and satisfies the condition $x(t_0) = x_0$.

Proof. Let $[a, b]$ be a closed interval contained in J and containing t_0. Let $t_0 \neq a, b$. Then we can show the existence of a unique solution $x(t)$ of the system (S_f) defined on $[a, b]$ and such that $x(t_0) = x_0$. In fact, consider first the interval $[a, t_0]$ and the operator T defined on $C_n[a, t_0]$ as follows:

$$(Tx)(t) = x_0 + \int_{t_0}^{t} [A(s)x(s) + f(s)]ds, \quad t \in [a, t_0]$$

In order to apply the contraction principle, we modify the norm of $C_n[a, t_0]$. We use the so called Bielecki norm (see also Example 2.2) which is defined, for $u \in C_n[a, t_0]$, to be

$$\|u\|_e = \sup_{a \le t \le t_0} \exp[-k(t_0 - t)] \|u\|$$

Here k is a fixed positive constant with

$$k > \max_{a \le t \le t_0} \|A(t)\| = r$$

$C_n[a, t_0]$ is a Banach space with the above norm. We now have

$$\|(Tu_1)(t) - (Tu_2)(t)\| \le \int_{t}^{t_0} \|A(s)(u_1(s) - u_2(s))\| ds$$

$$\leq \int_t^{t_0} \|A(s)\| \, \|u_1(s) - u_2(s)\| \, ds$$

$$\leq r \int_t^{t_0} \|u_1(s) - u_2(s)\| \, ds$$

$$\leq r \|u_1 - u_2\|_e \int_t^{t_0} e^{k(t_0-s)} \, ds$$

If we multiply the above by $e^{-k(t_0-t)}$, we obtain

$$e^{-k(t_0-t)} \|(Tu_1)(t) - (Tu_2)(t)\| \leq r \|u_1 - u_2\|_e \int_t^{t_0} e^{k(t-s)} \, ds$$

$$\leq (r/k) \|u_1 - u_2\|_e$$

Thus,

$$\|Tu_1 - Tu_2\|_e \leq (r/k) \|u_1 - u_2\|_e, \quad u_1, u_2 \in C_n[a, t_0]$$

Since $r/k < 1$, T is a contraction on $C_n[a, t_0]$, and an application of the contraction principle (Theorem 2.1) yields a unique fixed point $u(t)$ of T, which is the unique solution of (S_f) on $[a, t_0]$. Similarly one works on the interval $[t_0, b]$ to obtain the unique solution $\bar{u}(t)$ of (S_f) on this internal with the property $\bar{u}(t_0) = x_0$. Joining these two solutions we obtain the unique solution $x(t)$ of (S_f) on the interval $[a, b]$. Naturally, if $t_0 = a$ or $t_0 = b$, one of these two solutions equals $x(t)$ on $[a, b]$. Now we prove the theorem in the case $J = (-\infty, b)$. All the other cases, finite or infinite, can be handled similarly. Suppose that $x(t)$ is a solution of (S_f) with $x(t_0) = x_0$ which cannot be continued to $-\infty$. Then $x(t)$ is defined on a largest interval (c, d). Applying the above considerations of existence on closed intervals, we can show that (S_f) has a unique solution $\bar{x}(t)$ on the interval $[c, t_0]$ such that $\bar{x}(t_0) = x_0$. Obviously, $\bar{x}(t) = x(t)$, $t \in (c, t_0]$, which proves that $x(t)$ is continuable to the point $t = c$; that is, a contradiction. A similar argument applies to the case $d < b$. Thus, $x(t)$ is the unique solution of (S_f) on $J = (-\infty, b)$. This completes the proof.

The method of proof of the above theorem can now be applied to obtain a more general result.

Theorem 3.11 Let $F: J \times R^n \to R^n$ be continuous, where J is an interval of R. Assume further that for every interval $[a, b] \subset J$ there exists a constant $k > 0$ depending on $[a, b]$ and such that

$$\|F(t, u_1) - F(t, u_2)\| \leq k \|u_1 - u_2\|$$

for every $t \in [a, b]$, $u_1, u_2 \in R^n$. Then if $(t_0, x_0) \in J \times R^n$, System (3.1) possesses a unique solution $x(t)$ defined on J and such that $x(t_0) = x_0$.

We now consider the matrix system

$$X' = A(t)X, \quad t \in J \tag{S_A}$$

where J is an interval of R and $A: J \to M_n$ is continuous. Here we seek a solution $X: J \to M_n$. The existence and uniqueness theory of systems of the form (S_A) is identical to the corresponding theory of systems of the form (S). Thus, we have Theorem 3.12.

Theorem 3.12 Consider (S_A) with $A: J \to M_n$ continuous. Fix $t_0 \in J$ and $B \in M_n$. Then there exists a unique solution $X(t)$ of (S_A) which is defined on J and satisifes $X(t_0) = B$.

The following theorem characterizes the space of solutions of (S).

Theorem 3.13 Let $A: J \to M_n$. Then the set of all solutions of (S) is an n-dimensional vector space; that is, there exists a set P of n linearly independent solutions of (S) such that every other solution of (S) is a linear combination of the solutions in P.

Proof. Obviously, the set of all solutions of (S) is closed under addition and scalar multiplication (by real scalars). Thus it is a vector space. Let us first show that there exist at least n linearly independent solutions of (S). To this end, let $\bar{x}_i$, $i = 1, 2, \ldots, n$, denote the n vectors (in R^n)

$$\bar{x}_1 = \begin{bmatrix} 1 \\ 0 \\ \vdots \\ 0 \end{bmatrix}, \quad \bar{x}_2 = \begin{bmatrix} 0 \\ 1 \\ \vdots \\ 0 \end{bmatrix}, \ldots, \bar{x}_n = \begin{bmatrix} 0 \\ 0 \\ \vdots \\ 1 \end{bmatrix}$$

respectively. Then let t_0 be a point in J and consider the solutions $x_i(t)$, $i = 1, 2, \ldots, n$, of (S) which satisfy $x_i(t_0) = \bar{x}_i$. These solutions exist on J and are unique by Theorem 3.10. Now assume that the set $\{x_i(t); i = 1, 2, \ldots, n\}$ is linearly dependent on J. Then there are constants c_i, $i = 1, 2, \ldots, n$, not all zero, such that

$$\sum_{i=1}^{n} c_i x_i(t) = 0, \quad t \in J$$

In particular,

$$\sum_{i=1}^{n} c_i \bar{x}_i = \sum_{i=1}^{n} c_i x_i(t_0) = 0$$

This, however, is a contradiction, because the vectors $\bar{x}_i$, $i = 1, 2, \ldots, n$, are linearly independent. Now let $\bar{x}(t)$ be any solution of (S) and consider the algebraic system (in c_i)

$$\sum_{i=1}^{n} c_i \bar{x}_i = \sum_{i=1}^{n} c_i x_i(t_0) = \bar{x}(t_0)$$

This system has a unique solution, say $\bar{c}_1, \bar{c}_2, \ldots, \bar{c}_n$, because the determinant of its coefficients equals 1. Define the function $y(t)$ as follows:

$$y(t) = \sum_{i=1}^{n} \bar{c}_i x_i(t), \quad t \in J$$

Then $y(t)$ satisfies (S) on J and $y(t_0) = \bar{x}(t_0)$. Since the solutions of (S) are unique, $y(t) = \bar{x}(t)$, $t \in J$, which shows that $\bar{x}(t)$ is a linear combination of the functions $x_i(t)$. This completes the proof.

Definition 3.14 Consider System (S) with $A: J \to M_n$ continuous. Let $x_i(t)$, $i = 1, 2, \ldots, n$, be any linearly independent solutions of (S). Then the matrix $X(t)$, $t \in J$, whose columns are the n solutions $x_i(t)$, is called a "fundamental matrix of (S)."

The connection between the systems (S), (S_A) is established in Theorem 3.15.

Theorem 3.15 Let $A: J \to M_n$ be continuous. Then every fundamental matrix $X(t)$ of (S) satisfies the matrix system (S_A). Moreover, if $X(t_0) = B$ for some $t_0 \in J$ and $B \in M_n$, then $X(t)$ is the unique solution of (S_A) having the value B at t_0.

Proof. It is easy to see that

$$X'(t) = A(t)X(t), \quad t \in J$$

The rest of the proof follows from Theorem 3.12.

Now, under the assumptions of Theorem 3.15, let $X(t)$, $t \in J$, be a fundamental matrix of (S). Then it is easy to see that $x(t) = X(t)x_0$ is a solution of (S) for any $x_0 \in R^n$. Moreover, $x(t_0) = X(t_0)x_0$. Thus, if we show that $X^{-1}(t_0)$ exists for any fundamental matrix $X(t)$ of (S), then the solution $x(t)$ of (S) with $x(t_0) = u_0$, for some $(t_0, u_0) \in J \times R^n$, will be given by $x(t) = X(t)X^{-1}(t_0)u_0$. To this end, assume $X(t_0)$ is singular for some $t_0 \in J$. Then the equation $X(t_0)y = 0$ has a nonzero solution $y_0 \in R^n$. But this implies that the function $y(t) = X(t)y_0$, $t \in J$, is a solution of (S) with $y(t_0) = X(t_0)y_0 = 0$. Since the solutions of (S) are unique, we must have

$y(t) = X(t)y_0 \equiv 0$, $t \in J$. This implies that the columns of $X(t)$ are linearly dependent on J; that is, a contradiction. Therefore, $X^{-1}(t_0)$ exists.

The following theorem summarizes the above and gives an expression for the general solution of (S_f) in terms of a fundamental matrix $X(t)$ and the function $f(t)$.

Theorem 3.16 Let $A: J \to M^n$, $f: J \to R^n$ be continuous. Let $(t_0, x_0) \in J \times R^n$ be given. Then the unique solution $u(t)$ of the linear system (S), such that $u(t_0) = x_0$, is given by $u(t) = X(t)X^{-1}(t_0)x_0$, where $X(t)$ is any fundamental matrix of (S). Furthermore, the unique solution $x(t)$ of (S_f), such that $x(t_0) = x_0$, is given by the formula

$$x(t) = X(t)X^{-1}(t_0)x_0 + X(t)\int_{t_0}^{t} X^{-1}(s)f(s)ds \qquad (3.12)$$

where, again, $X(t)$ is any fundamental matrix of (S).

Proof. Since $x(t_0) = x_0$, it suffices to show that the function

$$v(t) = X(t)\int_{t_0}^{t} X^{-1}(s)f(s)ds$$

is a particular solution of (S_f). In fact,

$$v'(t) = X'(t)\int_{t_0}^{t} X^{-1}(s)f(s)ds + X(t)X^{-1}(t)f(t)$$
$$= A(t)X(t)\int_{t_0}^{t} X^{-1}(s)f(s)ds + f(t)$$
$$= A(t)v(t) + f(t)$$

This completes the proof.

Equation (3.12) is called the "variation of constants formula" for the system (S_f).

For a constant matrix $A \in M_n$ we have

Theorem 3.17 Let $A \in M_n$ be given. Then the fundamental matrix $X(t)$, $t \in R$, of the system

$$x' = Ax \qquad (3.13)$$

with the property $X(0) = I$, is given by the formula $X(t) = e^{tA}$. Moreover, the variation of constants formula (3.12) becomes now

$$x(t) = \exp[(t - t_0)A]x_0 + \int_{t_0}^{t} \exp[(t - s)A]f(s)ds \qquad (3.14)$$

Proof. Let $X(t) = e^{tA}$. Then, for $h \neq 0$, we have

$$X(t + h) - X(t) = e^{(t+h)A} - e^{tA} = (e^{hA} - I)e^{tA}$$

However,

$$e^{hA} - I = hA + \frac{(hA)^2}{2!} + \cdots$$

$$= hA + hL(hA)$$

where $L(hA) \to 0$ as $h \to 0$. Consequently,

$$\lim_{h \to 0} \frac{X(t + h) - X(t)}{h} = X'(t) = Ae^{tA}$$

Since (3.13) has unique solutions w.r.t. initial conditions, the first part of the proof is complete. The second part follows from $x(t_0) = x_0$ and one differentiation of (3.14).

EXERCISES

3.1. Prove Lemma 3.4.

3.2. Let $F: [a, b] \times R^n \to R^n$ be continuous and satisfy the following Lipschitz condition:

$$\|F(t, u_1) - F(t, u_2)\| \leq L\|u_1 - u_2\|, \quad t \in [a, b]$$

for any $u_1, u_2 \in R^n$ and some positive constant L. Use Gronwall's inequality to show that the problem

$$x' = F(t, x), \quad x(a) = x_0$$

can have at most one solution on $[a, b]$.

3.3. (Continuity w.r.t. initial conditions). Let F be as in Exercise 3.2. Moreover, let $\bar{x}_m \in R^n$ converge to $\bar{x} \in R^n$ as $m \to \infty$. Show that the solution $x_m(t)$, $t \in [a, b]$, of the problem

$$x' = F(t, x), \quad x(a) = \bar{x}_m$$

converges as $m \to \infty$ to the unique solution $x(t)$ of the problem

$$x' = F(t, x), \quad x(a) = \bar{x}$$

uniformly on $[a, b]$.

3.4. Let $A: J \to M_n$ (J is a subinterval of R) be continuous. Let $t_0 \in J$ be given and $X(t)$ be the fundamental matrix of the system (S) with $X(t_0) = I$. Furthermore, let $X_1(t)$ denote the fundamental matrix of the system $x' = -A^T x$ (called the "adjoint system") with $X_1(t_0) = I$. Show that $X_1^T(t) = X^{-1}(t)$ for every $t \in J$. Moreover, if $-A^T = A$, show that $\|y(t)\| = $ constant for any solution $y(t)$ of (S).

3.5. Let $A(t)$ be as in the first part of Exercise 3.4 and let $X(t)$ be a fundamental matrix of (S). Let $Y(t)$ be any other fundamental matrix of (S). Show that there exists a constant nonsingular matrix B such that $X(t)B = Y(t)$, $t \in J$.

3.6. (Liouville's formula). Let $A: [a, b] \to M_n$ be continuous. Let $X(t)$ be a fundamental matrix of (S) and $t_0 \in [a, b]$. Then

$$|X(t)| = |X(t_0)| \exp\left[\int_{t_0}^t trA(s)ds\right], \quad t \in [a, b]$$

where $|\cdot|$ denotes determinant and trA denotes the trace of the matrix A. [Hint. Show that $u(t) = |X(t)|$ satisfies the differential equation $u' = trA(t)u$.]

3.7. Let $A: R \to M_n$ be continuous and such that $A(t + L) = A(t)$ for $t \in R$, where L is some positive constant. Show that in order to obtain a solution $x(t)$ of (S) with $x(t + L) = x(t)$, $t \in R$, it suffices to show the existence of $x(t)$, $t \in [0, L]$, with $x(0) = x(L)$. Furthermore, show that $x(t) \equiv 0$ is the only such solution if and only if $I - X(L)$ is nonsingular. Here $X(t)$ is the fundamental matrix of (S) with $X(0) = I$.

3.8. Let $A: R_+ \to M_n$, $f: R_+ \to R^n$ be continuous and such that

$$\int_0^\infty \|A(t)\| dt < +\infty, \quad \int_0^\infty \|f(t)\| dt < +\infty$$

Using Gronwall's inequality, show that every solution $x(t)$ of (S_f) is bounded, that is, there exists $K > 0$ (depending on the solution $x(t)$) such that $\|x(t)\| \leq K$, $t \in R_+$. Moreover, show that every solution of (S_f) belongs to C_n^f.

3.9. Let $A: R_+ \to M_n$ be continuous and such that $\|A(t)\| \le K$, $t \ge 0$, where K is a positive constant. Prove that any solution $x(t) \ne 0$ of (S) satisfies

$$\limsup_{t \to \infty} \frac{\ln \|x(t)\|}{t} < +\infty$$

3.10. Consider the scalar problem

$$x' = |x|^\beta, \quad x(0) = 0$$

where $\beta \in (0, 1)$ is a constant. Show that every function of the type

$$x(t) = \begin{cases} 0, & t \le c \\ [(1-\beta)(t-c)]^{1/(1-\beta)}, & t \ge c \end{cases}$$

is a solution to this problem. Here c is any positive constant. Conclude that $f(x) = |x|^\beta$ cannot satisfy a Lipschitz condition on any interval containing zero.

3.11. Consider the scalar problem

$$x' = f(x), \quad x(0) = 0, \quad t \ge 0$$

where $f: R \to R_+$ is continuous, $f(x) > 0$ for $x \ne 0$, and $f(0) = 0$. Show that this problem has an infinity of solutions if

$$\int_{0^+}^{\epsilon} \frac{du}{f(u)} < +\infty$$

for some $\epsilon > 0$. If the above integral equals $+\infty$, then the only solution to the problem is the zero solution.

3.12. For the system (3.1), assume that $F: R_+ \times R^n \to R^n$ is continuous and such that

$$\|F(t, x)\| \le p(t)\|x\| + q(t)$$

where $p, q: R_+ \to R_+$ are continuous. Apply Gronwall's inequality and Theorem 3.8 to conclude that all local solutions of (3.1) are continuable to $+\infty$.

3.13. Using Exercise 3.12, discuss the continuation to the right of the local solutions of the system

$$\begin{bmatrix} x_1 \\ x_2 \\ x_3 \end{bmatrix}' = \begin{bmatrix} t\sin x_1 \\ (\cos t)\ln(|x_2| + 1) \\ e^{-t}x_3 \end{bmatrix} + \begin{bmatrix} t^2 \\ \sin t \\ e^{2t} \end{bmatrix}$$

3.14. The scalar equation

$$y'' + y = f(t)$$

can be written in system form as follows:

$$x_1' = x_2$$

$$x_2' = -x_1 + f(t)$$

Here $x_1 = y$. Using the variation of constants formula, express the general solution of this system in terms of its initial condition $x(0) = x_0$, the fundamental matrix $X(t)$ ($X(0) = I$) of the linear system

$$\begin{bmatrix} x_1 \\ x_2 \end{bmatrix}' = \begin{bmatrix} 0 & 1 \\ -1 & 0 \end{bmatrix} \begin{bmatrix} x_1 \\ x_2 \end{bmatrix}$$

and the function f.

3.15. Examine the local existence, uniqueness and extendability to the right of solutions of the scalar equation

$$x' = |x|^\beta$$

where $\beta > 1$ is a constant. Compare this situation to that of Exercise 3.10 for the initial condition $x(0) = 0$.

3.16. Solve the scalar problem

$$x' = x^3/t^{1/3}, \quad x(0) = 2$$

in the spirit of Theorem 3.3.

3.17. Solve the problem

$$\begin{bmatrix} x_1 \\ x_2 \end{bmatrix}' = \begin{bmatrix} 0 & 1 \\ 1 & 0 \end{bmatrix} \begin{bmatrix} x_1 \\ x_2 \end{bmatrix} + \begin{bmatrix} 0 \\ \sin t \end{bmatrix}, \quad \begin{bmatrix} x_1(0) \\ x_2(0) \end{bmatrix} = \begin{bmatrix} 1 \\ 0 \end{bmatrix}$$

by the use of the variation of constants formula.

3.18. Let the assumptions of Exercise 3.2 be satisfied with $a = 0$. Using Gronwall's inequality, show that the problem

$$x' = F(t, x), \ x(b) = x_1$$

has a unique solution defined on the interval $[0, b]$. Provide and prove a statement concerning continuity w.r.t. "final" conditions $x_n(b) = x_n$ analogous to that of Exercise 3.3.

3.19. Using Theorem 3.8, show that every solution to the scalar problem

$$x' = (\sin x)(1 - x^2)^{1/2}, \ x(0) = 1/2$$

is continuable to the point $t = 1/3$.

CHAPTER FOUR

STABILITY OF LINEAR SYSTEMS; PERTURBED LINEAR SYSTEMS

In this chapter we study the stability of systems of the form

$$x' = F(t, x) \tag{E}$$

where $F: R_+ \times R^n \to R^n$ is continuous.

A solution $x_0(t)$, $t \in R_+$, of the system (E) is stable, if the solutions of (E) which start close to $x_0(t)$ at the origin, remain close to $x_0(t)$ for all $t \in R_+$ in a certain sense. This actually means that small disturbances in the system that effect small perturbations to the initial conditions of solutions close to $x_0(0)$ do not really cause a considerable change to these solutions over the interval R_+. The various concepts of stability that we study in this chapter are actually dealing with the fashion in which the solutions close to $x_0(t)$ initially behave on infinite subintervals of R.

Although there are numerous types of stability, we present here only five types that are most important in the applications of linear and perturbed linear systems.

Other stability results can be found in Chapters 5, 7, and 9.

1. Definitions of Stability

In the following definitions $x_0(t)$ will denote a fixed solution of (E) defined on $[0, \infty)$.

Definition 4.1 The solution $x_0(t)$ is called "stable" if for every $\epsilon > 0$ there exists $\delta(\epsilon) > 0$ such that every solution $x(t)$ of (E) with $\|x(0) - x_0(0)\| < \delta(\epsilon)$ exists and satisfies $\|x(t) - x_0(t)\| < \epsilon$ on R_+. The solution $x_0(t)$ is called "asymptotically stable" if it is stable and there exists a constant $\eta > 0$ such that $x(t) - x_0(t) \to 0$ as $t \to \infty$ whenever $\|x(0) - x_0(0)\| \le \eta$. The solution $x_0(t)$ is called "unstable" if it is not stable.

Definition 4.2 The solution $x_0(t)$ is called "uniformly stable" if for every $\epsilon > 0$ there exists $\delta(\epsilon) > 0$ such that every solution $x(t)$ of (E) with $\|x(t_0) - x_0(t_0)\| < \delta(\epsilon)$ for some $t_0 \ge 0$ exists and satisfies $\|x(t) - x_0(t)\| < \epsilon$ on $[t_0, \infty)$. It is called "uniformly asymptotically stable" if it is uniformly stable and there exists $\eta > 0$ with the propery: for every $\epsilon > 0$ there exists $T(\epsilon) > 0$ such that $\|x(t_0) - x_0(t_0)\| < \eta$ for some $t_0 \ge 0$ implies $\|x(t) - x_0(t)\| < \epsilon$ for every $t \ge t_0 + T(\epsilon)$.

It is obvious that uniform stability implies stability and that uniform asymptotic stability implies asymptotic stability.

Definition 4.3 The solution $x_0(t)$ is called "strongly stable" if for every $\epsilon > 0$ there exists $\delta(\epsilon) > 0$ such that every solution $x(t)$ of (E) with $\|x(t_0) - x_0(t_0)\| < \delta(\epsilon)$ for some $t_0 \ge 0$ exists and satisfies $\|x(t) - x_0(t)\| < \epsilon$ on R_+.

Naturally, strong stability implies uniform stability. We should mention here that the interval $[0, \infty)$ in the definitions of stability can be replaced by any (but fixed) interval $[t_1, \infty)$ of the real line. We should also mention that $x_0(t)$ can be considered to be the zero solution. In fact, if $x(t) \equiv 0$ is not a solution of (E), then the transformation $u(t) = x(t) - y(t)$, where $y(t)$ is a fixed solution of (E), takes (E) into the system.

$$u' = F(t, u + y(t)) - F(t, y(t))$$

$$= G(t, u) \qquad (4.1)$$

This system has the function $u(t) \equiv 0$ as a solution. The stability properties of this solution correspond to the stability properties of the solution $y(t)$.

2. Linear Systems

In this section we study the stability properties of the linear systems

PERTURBED LINEAR SYSTEMS   61

$$x' = A(t)x, \tag{S}$$

$$x' = A(t)x + f(t) \tag{S_f}$$

where $A: R_+ \to M_n$, $f: R_+ \to R^n$ are continuous. It is clear that the solution $x_0(t)$ of (S_f) satisfies one of the definitions of stability of the previous section if and only if the zero solution of (S) has the same property. This follows from the fact that stability involves differences of solutions, combined with the superposition principle. Consequently, we may talk about the stability of (S_f) instead of the stability of one of its particular solutions. This will be done in the sequel even if $f \equiv 0$.

Theorem 4.4 Let $X(t)$ be a fundamental matrix of (S). Then (S) is stable if and only if there exists a constant $K > 0$ with

$$\|X(t)\| \le K, \ t \in R_+ \tag{4.2}$$

The system (S) is asymptotically stable if and only if

$$\|X(t)\| \to 0 \text{ as } t \to \infty \tag{4.3}$$

The system (S) is uniformly stable if and only if there exists a constant $K > 0$ with

$$\|X(t)X^{-1}(s)\| \le K, \ 0 \le s \le t < +\infty \tag{4.4}$$

The system (S) is uniformly asymptotically stable if and only if there exist constants $\alpha > 0$, $K > 0$ with

$$\|X(t)X^{-1}(s)\| \le Ke^{-\alpha(t-s)}, \ 0 \le s \le t < +\infty \tag{4.5}$$

The system (S) is strongly stable if and only if there exists a constant $K > 0$ such that

$$\|X(t)\| \le K, \ \|X^{-1}(t)\| \le K, \ t \in R_+ \tag{4.6}$$

Proof. We may assume that $X(0) = I$ because the conditions in the hypotheses hold for any fundamental matrix of (S) if they hold for a particular one. Assume first that (4.2) holds, and let $x(t)$, $t \in R_+$, be a solution of (S) with $x(0) = x_0$. Then, since $x(t) = X(t)x_0$, if for a given $\epsilon > 0$ we choose $\delta(\epsilon) = K^{-1}\epsilon$, we have

$$\|x(t)\| = \|X(t)x_0\| < \epsilon \text{ for } \|x_0\| < \delta(\epsilon) \tag{4.7}$$

Thus, System (S) is stable. Conversely, suppose that (S) is stable and fix $\epsilon > 0$, $\delta(\epsilon) > 0$ with the property

$$\|X(t)x_0\| < \epsilon$$

for every $x_0 \in R^n$ with $\|x_0\| < \delta(\epsilon)$. For a fixed $t \in R_+$ we get

$$[1/\delta(\epsilon)] \|X(t)x_0\| = \|X(t)(x_0/\delta(\epsilon))\| < \epsilon/\delta(\epsilon) \tag{4.8}$$

Since $x_0/\delta(\epsilon)$ ranges over the interior of the unit ball, we obtain

$$\|X(t)\| = \sup_{\|u\|<1} \|X(t)u\| \leq \epsilon/\delta(\epsilon) \tag{4.9}$$

This completes the proof of the first case because (4.9) holds for arbitrary $t \in R_+$.

Now assume that (4.3) holds. Then (4.2) holds for some $K > 0$ and

$$\lim_{t \to \infty} x(t) = \lim_{t \to \infty} X(t)x_0 = 0$$

for any solution $x(t)$ of (S) with $x(0) = x_0$. Thus (S) is asymptotically stable. Conversely, assume that (S) is asymptotically stable. Then there exists $\eta > 0$ such that $X(t)x_0 \to 0$ for every $x_0 \in R^n$ with $\|x_0\| \leq \eta$. Choose

$$x_0 = \begin{bmatrix} \eta \\ 0 \\ \cdot \\ \cdot \\ \cdot \\ 0 \end{bmatrix}$$

Since $X(t)x_0 = \eta y(t)$, where $y(t)$ is the first column of $X(t)$, we obtain that every entry of the first column of $X(t)$ tends to zero as $t \to \infty$. Similarly one concludes that every entry of $X(t)$ tends to zero as $t \to \infty$. This completes the proof of this case. In order to prove the third conclusion of the theorem, let (4.4) hold and let $t_0 \in R_+$ be given. Then $x(t) = X(t)X^{-1}(t_0)x_0$ is the solution of (S) with $x(t_0) = x_0$. Thus,

$$\|x(t)\| \leq \|X(t)X^{-1}(t_0)\| \|x_0\| \leq K \|x_0\|$$

for any $x_0 \in R^n$ with $\|x_0\| < K^{-1}\epsilon$, proves the uniform stability of S with $\delta(\epsilon) = K^{-1}\epsilon$. Now assume that (S) is uniformly stable. Fix $\epsilon > 0$, $\delta(\epsilon) > 0$ such that $\|X(t)X^{-1}(t_0)x_0\| < \epsilon$ for any $x_0 \in R^n$ with $\|x_0\| < \delta(\epsilon)$, any $t_0 \in R_+$ and any $t \geq t_0$. From this point on, the proof follows as the sufficiency part of the first case and is therefore omitted.

In the fourth case, let (4.5) hold. Then (S) is uniformly stable by virtue of (4.4). Now let ϵ ($0 < \epsilon < K$), $t_0 \in R_+$ be given, and let $x(t)$ be the solution of (S) with $\|x(t_0)\| = \|x_0\| < 1$. Then

$$\|x(t)\| = \|X(t)X^{-1}(t_0)x_0\| < K\exp\{-\alpha(t - t_0)\} \leq \epsilon$$

for every $t \geq t_0 + T(\epsilon)$, where $T(\epsilon) = -\alpha^{-1}\ln(\epsilon/K)$. Consequently, System (S) is uniformly asymptotically stable with the constant η of Definition 4.2 equal to 1. Conversely, let (S) be uniformly asymptotically stable. Fix ϵ ($0 < \epsilon < \eta$), $T = T(\epsilon)$, η as in Definition 4.3. Then $\|x_0\| = \eta$ implies

$$\|X(t)X^{-1}(t_0)x_0\| < \epsilon, \quad t \geq t_0 + T$$

Consequently, working as in the first case, we find

$$\|X(t)X^{-1}(t_0)\| \leq \mu < 1, \quad t \geq t_0 + T \tag{4.10}$$

where $\mu = \epsilon/\eta$. Now (4.4) implies the existence of a constant $K > 0$ such that

$$\|X(t)X^{-1}(t_0)\| \leq K, \quad t \geq t_0 \tag{4.11}$$

Thus, given $t \geq t_0$, there exists an integer $m \geq 0$ such that $t_0 + mT \leq t \leq t_0 + (m+1)T$. It follows that

$$\|X(t)X^{-1}(t_0)\| = \|X(t)X^{-1}(t_0 + mT)X(t_0 + mT)X^{-1}(t_0 + (m-1)T)X(t_0 +$$
$$(m-1)T)\cdots X^{-1}(t_0 + T)X(t_0 + T)X^{-1}(t_0)\|$$

$$\leq \|X(t)X^{-1}(t_0 + mT)\| \|X(t_0 + mT)X^{-1}(t_0 + (m-1)T)\|$$
$$\cdots \|X(t_0 + T)X^{-1}(t_0)\|$$

$$\leq K\mu^m$$

If we take $\alpha = -T^{-1}\ln\mu$, then

$$\|X(t)X^{-1}(t_0)\| \leq \mu^{-1}K\mu^{m+1} = \mu^{-1}K\mu^{-(m+1)\alpha T}$$

$$\leq \mu^{-1}Ke^{-\alpha(t-t_0)}$$

for every $t \geq t_0 + T$. This completes the proof of the case of uniform asymptotic stability.

Assume now that (4.6) holds and, given $\epsilon > 0$, choose $\delta(\epsilon) = K^{-2}\epsilon$. Then we have

$$\|X(t)X^{-1}(t_0)x_0\| \leq \|X(t)\| \|X^{-1}(t_0)\| \|x_0\|$$
$$\leq K^2 \|x_0\| < \epsilon$$

whenever $\|x_0\| < K^{-2}\epsilon$ and $t, t_0 \in R_+$. Thus, System (S) is strongly stable. To show the converse, let (S) be strongly stable and fix $\epsilon > 0$, $\delta(\epsilon) > 0$ such that

$$\|X(t)X^{-1}(t_0)x_0\| < \epsilon, \quad t, t_0 \in R_+ \tag{4.12}$$

whenever $\|x_0\| < \delta(\epsilon)$. This implies that for arbitrary $t, t_0 \in R_+$ we have

$$\|X(t)x_0\| < \epsilon, \quad \|X^{-1}(t_0)x_0\| < \epsilon$$

provided that $\|x_0\| < \delta(\epsilon)$. In fact, this follows from (4.12) if we take $t_0 = 0$, $t = 0$ respectively. Thus, as above,

$$\|X(t)\| \leq \epsilon/\delta(\epsilon), \quad \|X^{-1}(t)\| \leq \epsilon/\delta(\epsilon)$$

It follows that (4.6) holds for $K = \epsilon/\delta(\epsilon)$.

Before we consider System (S) with a constant matrix A, we should note that in the case of an "autonomous" system (that is, $F(t, x) \equiv F(x)$) stability is equivalent to uniform stability and asymptotic stability is equivalent to uniform asymptotic stability. This is a consequence of the fact that in this case $y(t) = x(t + \alpha)$ is a solution of (E) if $x(t)$ is a solution. This is true for any number $\alpha \in R$. Now consider the system

$$x' = Ax \tag{4.13}$$

with $A \in M_n$. If λ is an eigenvalue of A, then the dimension of the "eigenspace" of λ (the subspace of C^n generated by the eigenvectors of A corresponding to λ) is called the "index" of λ. The following theorem characterizes the fundamental matrices of (4.13) (cf. Cole [8, p. 89 and Theorem 4-7.1]).

Theorem 4.5 Let $X(t) = e^{tA}$ be the fundamental matrix of (4.13) with $X(0) = I$. Then every entry of $X(t)$ takes the form $e^{\alpha t}(p(t)\cos \beta t - q(t)\sin \beta t)$ or the form $e^{\alpha t}(p(t)\sin \beta t + q(t)\cos \beta t)$, where $\lambda = \alpha + \beta i$ is some eigenvalue of A and p, q are real polynomials in t. The degree d of the polynomial $p(t) + iq(t)$ satisfies $0 \leq d \leq m - r$, where m is the multiplicity of λ and r its index. Furthermore, if $m \neq r$, there is at least one entry of $X(t)$ such that $d \neq 0$.

Now we are ready to establish the stability properties of (4.13) in terms of the eigenvalues of the matrix A.

Theorem 4.6 The system (4.13) is stable and if and only if every eigenvalue of A that has multiplicity m equal to its index r has nonpositive real part, and every other eigenvalue has negative real part. The system (4.13) is asymptotically stable if and only if every eigenvalue of A has negative real part. It is strongly stable if and only if every eigenvalue of A is purely imaginary and has multiplicity equal to its index.

Proof. Let $X(t) = e^{tA}$, $t \in R_+$. Then (4.13) is stable if and only if $\|X(t)\| \leq K$, where $K > 0$ is a constant (cf. Theorem 4.4). Now let $\lambda = \alpha + \beta i$ be an eigenvalue of A. Then every entry of $X(t)$ corresponding to λ will be bounded if and only if $\alpha \leq 0$ for $m = r$ and $\alpha < 0$ for $m > r$. This completes the proof of our first assertion. The system (4.13) is asymptotically stable if and only if $e^{tA} \to 0$ as $t \to \infty$. This is of course possible if and only if every eigenvalue of A has negative real part. The system is strongly stable if and only if there exists a constant $K > 0$ such that $\|e^{tA}\| \leq K$, $\|e^{-tA}\| \leq K$ for every $t \in R_+$. Since e^{-tA} solves the system $X' = -AX$, and λ is an eigenvalue of A if and only if $-\lambda$ is an eigenvalue of $-A$, these inequalities can hold if and only if every eigenvalue of A has zero real part and $m = r$.

3. The Measure of a Matrix; Further Stability Criteria

Definition 4.7 Let A be an $n \times n$ matrix. Then $\mu(A)$ denotes the "measure of A" which is defined by

$$\mu(A) = \lim_{h \to 0^+} \frac{\|I + hA\| - 1}{h}$$

Theorem 4.8 The measure $\mu(A)$ exists as a finite number for every $A \in M_n$.

Proof. Let ϵ be given such that $0 < \epsilon < 1$, and consider the function

$$g(h) = \frac{\|I + hA\| - 1}{h}, \quad h > 0$$

Then we have

$$\|I + \epsilon h A\| = \|\epsilon(I + hA) + (1 - \epsilon)I\| \leq \epsilon \|I + hA\| + (1 - \epsilon)$$

or

$$g(\epsilon h) = \frac{\|I + \epsilon h A\| - 1}{\epsilon h} \leq \frac{\|I + hA\| - 1}{h} = g(h)$$

Thus, $g(h)$ is an increasing function of h. On the other hand,

$$\left| \frac{\|I + hA\| - 1}{h} \right| = \left| \frac{\|I + hA\| - \|I\|}{h} \right| \leq \frac{\|I + hA - I\|}{h} = \|A\|$$

This implies the existence of the limit of the function $g(h)$ as $h \to 0^+$. It follows that $\mu(A)$ exists and is finite.

Theorem 4.9 Let $A \in M_n$ be given. Then $\mu(A)$ has the following properties:

(i) $\mu(\alpha A) = \alpha \mu(A)$ for any $\alpha \in R_+$

(ii) $|\mu(A)| \leq \|A\|$

(iii) $\mu(A + B) \leq \mu(A) + \mu(B)$

(iv) $|\mu(A) - \mu(B)| \leq \|A - B\|$

Proof. Case (i) is trivial and (ii) follows from the fact that $|g(h)| \leq \|A\|$ for all $h > 0$, where g is as in the proof of Theorem 4.8. Inequality (iii) follows from

$$\frac{\|I + h(A + B)\| - 1}{h} \leq \frac{\|(\tfrac{1}{2})I + hA\| - (\tfrac{1}{2})}{h} + \frac{\|(\tfrac{1}{2})I + hB\| - (\tfrac{1}{2})}{h}$$

$$= \frac{\|I + 2hA\| - 1}{2h} + \frac{\|I + 2hB\| - 1}{2h}$$

Inequality (iv) follows easily from (ii) and (iii).

The following theorem establishes the relationship between the solutions of (S) and the measure of the matrix $A(t)$.

Theorem 4.10 Let $A: R_+ \to M_n$ be continuous. Then for every $t_0, t \in R_+$ with $t \geq t_0$ we have

$$\|x(t_0)\| \exp\left[-\int_{t_0}^{t} \mu(-A[s])ds\right] \leq \|x(t)\| \leq \|x(t_0)\| \exp\left[\int_{t_0}^{t} \mu(A(s))ds\right]$$

where $x(t)$ is any solution of (S).

Before we provide a proof of Theorem 4.10, we establish the auxiliary lemma 4.11.

Lemma 4.11 Let $r: [t_0, b) \to R_+$, $\phi: [t_0, b) \to R$ $(0 \le t_0 < b \le +\infty)$ be continuous and such that

$$r'_+(t) \le \phi(t) r(t), \quad t \in [t_0, b)$$

where r'_+ denotes the right derivative of the function $r(t)$. Then $r(t) \le u(t)$, $t \in [t_0, b)$, where $u(t)$ is the solution of

$$u' = \phi(t) u, \quad u(t_0) = r(t_0) \tag{4.14}$$

Proof. Let $t_1 \in (t_0, b)$ be an arbitrary point. We will show that $r(t) \le u(t)$ on the interval $[t_0, t_1]$. Consider first the solution $u_n(t)$, $t \in [t_0, t_1]$, of the problem

$$u' = \phi(t) u + 1/n, \quad u(t_0) = r(t_0), \quad n = 1, 2, \ldots \tag{4.15}_n$$

respectively. Fix n and assume the existence of a point $t_2 \in (t_0, t_1)$ such that $r(t_2) > u_n(t_2)$. Then there exists $t_3 \in [t_0, t_2)$ such that $r(t_3) = u_n(t_3)$ and $r(t) > u_n(t)$ on $(t_3, t_2]$. From $(4.15)_n$ we obtain that

$$u'_n(t_3) = \phi(t_3) u_n(t_3) + 1/n$$

$$= \phi(t_3) r(t_3) + 1/n$$

$$\ge r'_+(t_3) + 1/n$$

$$> r'_+(t_3)$$

Consequently, $u_n(t) > r(t)$ in a small right neighborhood of the point t_3. This is a contradiction to $r(t) > u_n(t)$ on $(t_3, t_2]$. Thus $r(t) \le u_n(t)$ for any $t \in [t_0, t_1]$ and any $n = 1, 2, \ldots$. Now we use Gronwall's inequality (Lemma 3.4) to show that $(4.15)_n$ actually implies

$$|u_n(t) - u_m(t)| \le 2 t_1 |1/n - 1/m| \exp \{ \int_{t_0}^{t_1} |\phi(s)| ds \}, \quad t \in [t_0, t_1] \tag{4.16}$$

for $n, m \ge 1$. Thus, the sequence $\{u_n(t)\}$, $n = 1, 2, \ldots$, is Cauchy. It follows that $u_n(t) \to u(t)$ as $n \to \infty$ uniformly on $[t_0, t_1]$, where $u(t)$ is the solution of the problem (4.14) on the interval $[t_0, b)$. Since t_1 is arbitrary, it follows that $r(t) \le u(t)$, $t \in [t_0, b)$.

It should be noted that a corresponding inequality holds if $r'_-(t) \ge \phi(t) r(t)$, where $r'_-(t)$ is the left derivative of $r(t)$ on (t_0, b).

Proof of Theorem 4.10. Let $r(t) = \|x(t)\|$. We are planning to show that

$$r'_+(t) \leq \mu(A(t))r(t) \tag{4.17}$$

To this end, we first notice that for any two vectors $x_1, x_2 \in R^n$ the limit

$$\lim_{h \to 0^+} \frac{\|x_1 + hx_2\| - \|x_1\|}{h}$$

exists as a finite number. To show this, it suffices to show that the function

$$g_1(h) = \frac{\|x_1 + hx_2\| - \|x_1\|}{h}$$

is increasing and bounded by $\|x_2\|$ on $(0, \infty)$. We omit the proof, which is very similar to the corresponding one for the function $g(h)$ in the proof of Theorem 4.8. It follows that the limit

$$\lim_{h \to 0^+} \frac{\|x(t) + hx'(t)\| - \|x(t)\|}{h} \tag{4.18}$$

exists as a finite number. We will show that this number equals $r'_+(t)$. In fact, let $h > 0$ be given. Then we have

$$\left| \frac{\|x(t+h)\| - \|x(t)\|}{h} - \frac{\|x(t) + hx'(t)\| - \|x(t)\|}{h} \right|$$

$$= \left| \frac{\|x(t+h)\| - \|x(t) + hx'(t)\|}{h} \right|$$

$$\leq \frac{\|x(t+h) - x(t) - hx'(t)\|}{h} \to 0 \text{ as } h \to 0^+$$

which proves that $r'_+(t)$ equals the limit in (4.18). Consequently,

$$r'_+(t) = \lim_{h \to 0^+} \frac{\|x(t) + hA(t)x(t)\| - \|x(t)\|}{h} \tag{4.19}$$

$$\leq \lim_{h \to 0^+} \frac{\|I + hA(t)\| - 1}{h} r(t)$$

$$\leq \mu(A(t))r(t)$$

Applying Lemma 4.11, we obtain

$$\|x(t)\| = r(t) \le \|x(t_0)\| \exp\{\int_{t_0}^{t} \mu(A(s))ds\} \quad (4.20)$$

for every $t \ge t_0$. In order to find the corresponding lower bound of $\|x(t)\|$, let $u = -t$, $u_0 = -t_0$. Then $y(u) = x(-u)$, $u \in (-\infty, u_0]$, satisfies the system

$$y' = -A(-u)y$$

Thus, as in (4.20), we get

$$\|y(u_0)\| \le \|y(u)\| \exp\{\int_{u}^{u_0} \mu(-A(-s))ds\}$$

$$= \|y(u)\| \exp\{-\int_{-u}^{-u_0} \mu(-A(v))dv\}, \quad u_0 \ge u$$

or

$$\|x(t_0)\| \le \|x(t)\| \exp\{\int_{t_0}^{t} \mu(-A(s))ds\}, \quad t \ge t_0$$

This completes the proof.

We are now ready for the main theorem of this section.

Theorem 4.12 Consider System (S) with $A: R_+ \to M_n$ continuous. If

$$\liminf_{t \to \infty} \int_0^t \mu(-A(s))ds = -\infty$$

then (S) is unstable. If

$$\limsup_{t \to \infty} \int_0^t \mu(A(s))ds < +\infty$$

then (S) is stable. If

$$\lim_{t \to \infty} \int_0^t \mu(A(s))ds = -\infty$$

then (S) is asymptotically stable. If

$$\mu(A(t)) \le 0, \quad t \ge 0$$

then (S) is uniformly stable. If, for some $r > 0$,

$$\mu(A(t)) \le -r, \quad t \ge 0$$

then (S) is uniformly asymptotically stable.

The proof is left as an exercise.

The following table provides formulas for $\mu(A)$ corresponding to the three different norms of Example 1.5.

Table 4.13

$\|x\|$	$\mu(A)$
$\|x\|_1$	largest eigenvalue of $\tfrac{1}{2}(A + A^T)$
$\|x\|_2$	$\max_i \{a_{ii} + \sum_{j,\, j \ne i} |a_{ij}|\}$
$\|x\|_3$	$\max_j \{a_{jj} + \sum_{i,\, i \ne j} |a_{ij}|\}$

4. Perturbed Linear Systems

In this section we study the stability of systems of the form

$$x' = A(t)x + F(t, x) \qquad (S_F)$$

where $A: R_+ \to M_n$, $F: R_+ \times R^n \to R^n$ are continuous functions with $F(t, 0) \equiv 0$, $t \in R_+$. We start with a theorem concerning the asymptotic stability of (S_F). The proof of this theorem is based on Lemma 4.13.

Lemma 4.13 Let $X(t)$ be a fundamental matrix of the system (S). Assume further that there exists a constant $K > 0$ such that

$$\int_0^t \|X(t)X^{-1}(s)\| \, ds \le K, \quad t \ge 0 \qquad (4.21)$$

Then there exists a constant $M > 0$ such that

$$\|X(t)\| \le M e^{-K^{-1}t}, \quad t \ge 0$$

Proof. Let $u(t) = \|X(t)\|^{-1}$. Then we have

$$\left(\int_0^t u(s) ds \right) X(t) = \int_0^t u(s) ds \, X(t) X^{-1}(s) X(s)$$

from which we obtain

$$\int_0^t u(s) ds \, \|X(t)\| \le \int_0^t \|X(t)X^{-1}(s)\| \, \|X(s)\| u(s) ds \le K, \quad t \ge 0$$

or

$$u(t) \geq K^{-1}\int_0^t u(s)ds \tag{4.22}$$

Now let $\lambda(t)$ denote the integral on the right hand side of (4.22). Then we have

$$\lambda'(t) \geq K^{-1}\lambda(t), \quad t \geq 0 \tag{4.23}$$

Dividing (4.23) by $\lambda(t)$ and integrating from t_0 to $t > t_0$, we obtain

$$\lambda(t) \geq \lambda(t_0)e^{K^{-1}(t-t_0)}, \quad t \geq t_0$$

Consequently,

$$\|X(t)\| = [u(t)]^{-1} \leq K[\lambda(t)]^{-1} \leq [K/\lambda(t_0)]e^{-K^{-1}(t-t_0)}$$

for every $t \geq t_0$. We choose M so large that both

$$M \geq [K/\lambda(t_0)]e^{K^{-1}t_0}$$

and

$$\|X(t)\| \leq Me^{-K^{-1}t_0}, \quad 0 \leq t \leq t_0$$

This completes the proof.

Theorem 4.14 Let $X(t)$ be a fundamental matrix of (S) such that

$$\int_0^t \|X(t)X^{-1}(s)\|ds \leq K, \quad t \geq 0$$

Moreover, let

$$\|F(t, x)\| \leq \mu\|x\|, \quad t \geq 0$$

with μ satisfying $0 \leq \mu < K^{-1}$. Then the zero solution of (S_F) is asymptotically stable.

Proof. Let $X(t)$ be the fundamental matrix of (S) with $X(0) = I$. Then since $X(t)X^{-1}(s) = Y(t)Y^{-1}(s)$ for any other fundamental matrix $Y(t)$ of (S), Lemma 4.13 holds for this particular $X(t)$. Thus, $X(t) \to 0$ as $t \to \infty$. If $x(t)$ is a local solution of (S_F) defined to the right of $t = 0$, then $x(t)$ satisfies the system

$$u' = A(t)u + F(t, x(t))$$

Using the variation of constants formula for this system we obtain

CHAPTER FOUR

$$x(t) = X(t)x(0) + \int_0^t X(t)X^{-1}(s)F(s, x(s))ds \qquad (4.24)$$

Letting $L > 0$ be such that $\|X(t)\| \leq L$ for $t \geq 0$, we obtain

$$\|x(t)\| \leq L\|x(0)\| + \mu K \max_{0 \leq s \leq t} \|x(s)\|$$

which implies

$$\max_{0 \leq s \leq t} \|x(s)\| \leq (1 - \mu K)^{-1} L \|x(0)\|$$

It follows that

$$\|x(t)\| \leq (1-\mu K)^{-1} L \|x(0)\|$$

as long as $x(t)$ is defined. This implies that $x(t)$ is continuable to $+\infty$ (see Theorem 3.8), and that the zero solution is stable. Now we show that $x(t) \to 0$ as $t \to \infty$. To this end, let

$$c = \limsup_{t \to \infty} \|x(t)\|$$

and pick d such that $\mu K < d < 1$. If $c > 0$, then, since $d^{-1} c > c$, there exists $t_0 \geq 0$ such that

$$\|x(t)\| \leq d^{-1} c$$

for every $t \geq t_0$. Thus, (4.24) implies

$$\|x(t)\| = \|X(t)x(0) + X(t)\int_0^t X^{-1}(s)F(s, x(s))ds\|$$
$$= \|X(t)x(0) + X(t)\int_0^{t_0} X^{-1}(s)F(s, x(s))ds + X(t)\int_{t_0}^t X^{-1}(s)F(s, x(s))ds\|$$
$$\leq \|X(t)\|\|x(0)\| + \|X(t)\|\int_0^{t_0}\|X^{-1}(s)F(s, x(s))\|ds$$
$$\quad + \int_{t_0}^t \|X(t)X^{-1}(s)\|\|F(s, x(s))\|ds$$
$$\leq \|X(t)\|\|x(0)\| + \|X(t)\|\int_0^{t_0}\|X^{-1}(s)F(s, x(s))\|ds + \mu K d^{-1} c$$

Taking the lim sup above as $t \to \infty$, we obtain $c \leq \mu K d^{-1} c$; that is, a contradiction. Thus, $c = 0$. This completes the proof.

Theorem 4.15 has a corollary concerning the uniform stability of system (S_F).

Theorem 4.15 Let $X(t)$ be a fundamental matrix of (S) and let

$$\|X(t)X^{-1}(s)\| \le K, \quad t \ge s \ge 0$$

where K is a positive constant. Moreover, let

$$\|F(t, x)\| \le \lambda(t)\|x\|$$

where $\lambda: R_+ \to R_+$ is continuous and such that

$$\int_0^\infty \lambda(t)dt < +\infty$$

Then if

$$M = K \exp\left(K \int_0^\infty \lambda(t)dt\right)$$

every local solution $x(t)$ of (S_F), defined to the right of the point $t_0 \ge 0$, is continuable to $+\infty$ and satisfies

$$\|x(t)\| \le M\|x(t_0)\|$$

for every $t \ge t_0$.

Proof. From the variation of constants formula (4.24), with t_0 replacing 0, we have

$$\|x(t)\| \le K\|x(t_0)\| + K\int_{t_0}^t \lambda(s)\|x(s)\|ds$$

for $t \ge t_0$. Applying Gronwall's inequality (Lemma 3.4), we obtain

$$\|x(t)\| \le K\|x(t_0)\| \exp\left(K \int_{t_0}^t \lambda(s)ds\right) \le M\|x(t_0)\| \tag{4.25}$$

for $t \ge t_0$. Consequently, by Theorem 3.8, $x(t)$ is continuable to $+\infty$ and (4.25) holds for every $t \ge t_0$.

Corollary 4.16 If the system (S) is uniformly stable, and if F is as in Theorem 4.15, then the zero solution of (S_F) is uniformly stable. In particular, the uniform stability of (S) implies the uniform stability of the system

$$x' = [A(t) + B(t)]x \tag{4.26}$$

where $B: R_+ \to M_n$ is continuous and such that

$$\int_0^\infty \|B(t)\|dt < +\infty$$

The uniform asymptotic stability of the system (S_F) follows from Theorem 4.17.

Theorem 4.17 Let $X(t)$ be a fundamental matrix of (S) such that

$$\|X(t)X^{-1}(s)\| \leq Ke^{-\mu(t-s)}, \quad t \geq s \geq 0$$

where K, μ are positive constants. Let

$$\|F(t, x)\| \leq \lambda \|x\|$$

with λ, a positive constant, satisfying $\lambda < K^{-1}\mu$. Then if $c = \mu - \lambda K$, every solution $x(x)$ of (S_F), defined in a right neighborhod of t_0, exists for $t \geq t_0$ and satisfies

$$\|x(t)\| \leq Ke^{-c(t-s)} \|x(s)\|$$

for every t, s with $t \geq s \geq t_0$.

Proof. From the variation of constants formula

$$x(t) = X(t)X^{-1}(t_0)x(t_0) + \int_{t_0}^{t} X(t)X^{-1}(s)F(s, x(s))ds$$

in a right neighborhood of the point $t_0 \geq 0$, we obtain

$$\|x(t)\| \leq Ke^{-\mu(t-t_0)} \|x(t_0)\| + \lambda K \int_{t_0}^{t} e^{-\mu(t-s)} \|x(s)\| ds, \quad t \geq t_0$$

Consequently, if $z(t) = e^{\mu(t-t_0)} \|x(t)\|$, we have

$$z(t) \leq Kz(t_0) + \lambda K \int_{t_0}^{t} z(s)ds, \quad t \geq t_0$$

An application of Gronwall's inequality (Lemma 3.4) yields

$$z(t) \leq Kz(t_0)e^{\lambda K(t-t_0)}$$

for $t \geq t_0$, and

$$\|x(t)\| \leq K\|x(t_0)\| e^{-c(t-t_0)}$$

Obviously, $x(t)$ is continuable to $+\infty$ (see Theorem 3.8).

Corollary 4.18 If (S) is uniform asymptotically stable and if the constant λ of Theorem 4.17 is sufficiently small, then the zero solution of (S_F) is uniformly asymptotically stable. In particular, the uniform asymptotic stability of the system (S) implies the same property for the sytem (4.26), where $B: R_+ \to M_n$ is continuous and such that $B(t) \to 0$ as $t \to \infty$.

EXERCISES

4.1. Consider the scalar equation

$$x' = b(t)x \qquad (*)$$

where $b: R_+ \to R$ is continuous. Show that (*) is asymptotically stable if and only if

$$\int_0^\infty b(t)dt = -\infty$$

4.2. Prove Theorem 4.12.

4.3. Show that if

$$x' = A(t)x \qquad (S)$$

with $A: R_+ \to M_n$ continuous, is stable, and if

$$\int_0^\infty trA(s)ds \geq m, \quad t \geq 0$$

with m constant, then (S) is strongly stable.

4.4. Consider system (S_F) with $A: R_+ \to M_n$, $F: R_+ \times R^n \to R^n$ continuous. Furthermore, assume the existence of a positive constant c such that $\|F(t, x)\| \leq \lambda(t)\|x\|$ for every $x \in R^n$ with $\|x\| \leq c$, where $\lambda: R_+ \to R_+$ is continuous and such that

$$\int_0^\infty \lambda(t)dt = d < +\infty$$

Then the uniform stability of the homogeneous system (S), assuming (4.4), implies the stability of the zero solution of the system (S_F). [Hint. Show that if $\|x_0\| < \mu$, where

$$\mu = \min\{c, cK^{-1}e^{-dK}\}$$

then $\|x(t)\| < c$ as long as $x(t)$ exists. Here $x(t)$ is a solution of (S_F) with $x(0) = x_0$. To this end, assme that there is a point $\bar{t} \in (0, b)$ such that $\|x(\bar{t})\| = c$ and $\|x(t)\| < c$ on $[0, \bar{t})$. Then use the growth condition on f along with Gronwall's inequaltiy to obtain the contradiction: $\|x(t)\| \leq K\|x_0\|e^{dK} < c$ on $[0, \bar{t}]$. This last inequality on $[0, \infty)$ implies the stability of the zero solution of (S_F).]

4.5. Consider the scalar equation

$$x'' + x = (1 + t^3)^{-1}|x|^\alpha, \quad t \geq 0$$

where $\alpha > 1$ is a constant. Show that the system corresponding to this equation has its zero solution stable. [Hint. Apply the result of Exercise 4.4]

4.6. Consider the "autonomous" system $x' = F(x)$, where $F: R^n \to R^n$ is continuous and such that $F(0) = 0$. Show that the stability of the zero solution of this system is equivalent to its uniform stability.

4.7. How small should $|m|$ be (m constant) so that the zero solution of the equation

$$x'' + 2x' + x = [m/(1 + t^2)]\,|x|$$

written as a system, is asymptotically stable?

4.8. Let (S) be uniformly asymptotically stable. Let

$$\int_t^{t+1} \|f(s)\|\,ds \leq c, \quad t \geq 0 \qquad (*)$$

where $f: R_+ \to R^n$ is continuous and c is a positive constant. Show that every solution of (S_f) is bounded. [Hint. Show first that $(*)$ implies

$$e^{-\alpha t} \int_0^t e^{\alpha s}\|f(s)\|\,ds \leq c(1 - e^{-\alpha})^{-1}, \quad t \geq 0$$

for any $\alpha > 0$.]

4.9. Consider the system (S) with $A: R_+ \to M_n$ continuous. Let $\lim_{t \to \infty} A(t) = A_0$. Then, assuming "smallness" conditions on

$$\|A(t) - A_0\| \quad \text{or} \quad \int_0^\infty \|A(t) - A_0\|\,dt$$

obtain stability properties of (S) based on those of $x' = A_0 x$.

4.10. Let A_0 be as in Exercise 4.9 and let $A: R_+ \to M_n$, $f: R_+ \to R^n$ continuous. Show that if all solutions of $x' = A_0 x$ are bounded, then all solutions of (S_f) are bounded, provided that

$$\int_0^\infty \|A(t) - A_0\|\,dt < +\infty, \quad \int_0^\infty \|f(t)\|\,dt < +\infty$$

4.11. Consider (S_f) with $A: R_+ \to M_n$, $f: R_+ \to R^n$ continuous. Assume that the system (S) is uniformly asymptotically stable. Then if $\|f(t)\| \to 0$ as $t \to \infty$, every solution of (S_f) tends to zero as $t \to \infty$.

4.12. Let $A \in M_n$ and $F: R_+ \times R^n \to R^n$ be continuous in (S_f). Assume that the system (S) is asymptotically stable and that

$$\lim_{\|u\| \to 0} \|F(t, u)\|/\|u\| = 0, \quad \text{uniformly w.r.t } t \in R_+$$

Show that the zero solution of (S_F) is asymptotically stable.

4.13. Assume that $X(t)$ satisfies (4.6) and that F is as in Theorem 4.15. Let t_0 be an arbitrary point in R_+. Show that every solution $x(t)$ of (S_F) with sufficiently small initial value at t_0 is continuable to $+\infty$ and satisfies

$$\|x(t)\| \le M\|x(t_0)\|, \quad t \in R_+$$

where M is a positive constant independent of the solution $x(t)$. [Hint. Differentiate the function $q(t) = \|X^{-1}(t)x(t)\|$ from the left, by using

$$|q(t) - q(t-h)| \le K^2 \int_{t-h}^{t} \lambda(s) q(s) ds, \quad 0 < t - h < t$$

to obtain locally

$$-K^2 \lambda(t) q(t) \le q'_-(t) \le K^2 \lambda(t) q(t)$$

Use a suitable version of Lemma 4.11 to obtain integral inequalities as in Theorem 4.10. Then show that

$$\|X^{-1}(t)x(t)\| \le L\|X^{-1}(s)x(s)\|, \quad t, s \ge 0$$

where

$$L = \exp\{K^2 \int_0^\infty \lambda(t) dt\}]$$

4.14. Using Exercise 4.13, show that the strong stability of the system (S) implies the strong stability of (4.26), where $B: R_+ \to M_n$ is continuous and such that

$$\int_0^\infty \|B(t)\| dt < +\infty$$

CHAPTER FIVE

LYAPUNOV FUNCTIONS IN THE THEORY OF DIFFERENTIAL SYSTEMS; THE COMPARISON PRINCIPLE

Let $V: R^n \to R_+$ be a continuous function such that $\lim\limits_{\|x\| \to \infty} V(x) = +\infty$, and let $x(t)$ be a solution of the system

$$x' = F(t, x) \tag{E}$$

defined on R_+. In order to show that $x(t)$ is bounded on R_+, it suffices to show the boundedness of $V(x(t))$ on R_+. In fact, let $V(x(t)) \leq K$, $t \in R_+$, where K is a positive constant, and let $\|x(t_m)\| \to +\infty$ as $m \to \infty$, for some sequence $\{t_m\}_{m=1}^{\infty}$. Then $V(x(t_m)) \to +\infty$, which is a contradiction. It follows that $x(t)$ is bounded.

The main observation here is that local or global information about the solutions of (E) may be obtained from certain scalar functions that, in some sense, are associated with the system (E).

The study of this situation is the subject of this chapter.

In Section 1 we introduce the concept of such "Lyapunov functions" V and show their main connection with the system (E). Section 2 establishes the existence of "maximal" and "minimal" solutions of first order scalar problems

$$u' = \gamma(t, u), \ u(t_0) = u_0$$

and relationships with the system (E) by means of

$$V'_E(t, u) \leq \gamma(t, V(t, u)), \ V(t_0, u_0) \leq u_0$$

where V is a Lyapunov function associated with (E) and V'_E is a certain derivative of V "along the system (E)." This inequality, which provides information about $V(t, x(t))$, is the main ingredient of the so-called "Comparison Principle." Naturally, the scalar function $\gamma(t, u)$ above is intimately related to the system (E). Such a relation could be given, for example, by an inequaltiy of type

$$\|F(t, u)\| \leq \gamma(t, \|u\|)$$

on a subset of $R_+ \times R^n$.

An application of these considerations to an existence theorem on R_+ is given in Section 3, and Section 4 concerns itself with stability properties of the zero solution of (E).

1. Lyapunov Functions

Definition 5.1 Let S be a subset of $R_+ \times R^n$ and let $V: S \to R_+$ be continuous and satisfy a Lipschitz condition w.r.t. its second variable in every compact subset of S. This means that if $K \subset S$ is compact, then there exists a constant $L_K > 0$ such that

$$\|V(t, u_1) - V(t, u_2)\| \leq L_K \|u_1 - u_2\|$$

for every $(t, u_1), (t, u_2) \in K$. Then V is called a Lyapunov function.

In what follows, the function $F: R_+ \times R^n \to R^n$ in the equation (E) will be assumed continuous. Let $V(t, u)$ be a Lyapunov function defined on the **open** set S. We define V'_E as follows:

$$V'_E(t, u) = \lim_{h \to 0^+} \sup (1/h)[V(t + h, u + hF(t, u)) - V(t, u)] \qquad (5.1)$$

If $x(t)$ is a solution of (E) such that $(t, x(t)) \in S$, $t \in [a, b]$, $0 \leq a < b \leq +\infty$ then we define

$$V^+(t, x(t)) = \lim_{h \to 0^+} \sup (1/h) [V(t + h, x(t + h)) - V(t, x(t))] \qquad (5.2)$$

for every $t \in [a, b]$. If we fix t, then there exists a neighborhood N of the point $(t, x(t))$ such that $\bar{N} \subset S$, $(t + h, x(t) + hF(t, x(t))) \in N$ and $(t + h, x(t + h)) \in N$ for all sufficiently small $h > 0$. Then, for such h,

$$\frac{x(t + h) - x(t)}{h} - F(t, x(t)) = \epsilon(h)$$

where $\epsilon(h) \to 0$ as $h \to 0^+$. It follows that

$$V(t + h, x(t + h)) - V(t, x(t)) = V(t + h, x(t) + hF(t, x(t))$$
$$+ h\epsilon(h)) - V(t, x(t))$$
$$\leq V(t + h, x(t) + hF(t, x(t)))$$
$$+ Lh\|\epsilon(h)\| - V(t, x(t)) \quad (5.3)$$

where L is the (local) Lipschitz constant of V. In (5.3) we divide by $h > 0$ and take the lim sup as $h \to 0^+$ to obtain

$$V^+(t, x(t)) \leq V'_E(t, x(t)) \quad (5.4)$$

Similarly, we have

$$V(t + h, x(t + h)) - V(t, x(t)) \geq V(t + h, x(t) + hF(t, x(t)))$$
$$- Lh\|\epsilon(h)\| - V(t, x(t))$$

which yields

$$V'_E(t, x(t)) \leq V^+(t, x(t)) \quad (5.5)$$

Thus,

$$V'_E(t, x(t)) = V^+(t, x(t)) \quad (5.6)$$

at all points $t \in [a, b)$. Evidently, we also have

$$\liminf_{h \to 0^+} (1/h)[V(t + h, x(t + h)) - V(t, x(t))]$$
$$= \liminf_{h \to 0^+} (1/h)[V(t + h, x(t) + hF(t, x(t))) - V(t, x(t))] \quad (5.7)$$

We have shown the following lemma.

Lemma 5.2 Let S be an open subset of $R_+ \times R^n$ and let $(t, x(t)) \in S$ for every $t \in [a, b)$, $0 \leq a < b \leq +\infty$, where $x(t)$ is a solution of the system (E). Assume further that V is a Lyapunov function on S. Then we have

$$V'_E(t, x(t)) = V^+(t, x(t)) \quad (5.8)$$

for every $t \in [a, b)$.

In what follows, Lyapunov functions will be defined sometimes on sets S which are not necessarily open, but it will be obvious that some of the facts established above hold as well in such sets.

2. Maximal and Minimal Solutions; The Comparison Principle

The four Dini derivatives of a function $u(t)$ are defined as follows:

$$D^+ u(t) = \lim_{h \to 0^+} \sup (1/h)[u(t+h) - u(t)]$$

$$D_+ u(t) = \lim_{h \to 0^+} \inf (1/h)[u(t+h) - u(t)]$$

$$D^- u(t) = \lim_{h \to 0^-} \sup (1/h)[u(t+h) - u(t)]$$

$$D_- u(t) = \lim_{h \to 0^-} \inf (1/h)[u(t+h) - u(t)]$$

These functions are extended real-valued; that is, the values $+\infty$ and $-\infty$ are not excluded.

Definition 5.3 Consider the equation

$$u' = \gamma(t, u) \tag{5.9}$$

where γ is a continuous real valued function defined on a suitable subset of $R_+ \times R$. Let $u(t)$ be a solution of the equation (5.9) defined on an interval $I = [t_0, t_0 + a)$ $(0 < a \leq +\infty)$ or $I = [t_0, t_0 + a]$ with $t_0 \geq 0$. Then $u(t)$ is said to be the "maximal solution" of (E) on I, if for any other solution $v(t)$ of (E) with $v(t_0) = u(t_0)$ and domain $I_1 \subset I$, we have

$$u(t) \geq v(t), \ t \in I_1$$

Similarly one defines the "minimal solution."

Obviously the maximal and the minimal solutions on I are unique.

We are planning to show that there are always local maximal and minimal solutions of the system (E). Before we state the relevant result we need Theorem 5.4.

Theorem 5.4 Let $\gamma: \bar{M} \to R$ $(M \subset R^2$, open) be a continuous function. Let $v, w: [t_0, t_0 + a) \to R$ $(0 < a \leq +\infty)$ be continuous and such that $(t, v(t))$, $(t, w(t)) \in \bar{M}$ for all $t \in [t_0, t_0 + a)$. Furthermore, let $v(t_0) < w(t_0)$ and

$$D_- v(t) < \gamma(t, v(t))$$

$$D_- w(t) \geq \gamma(t, w(t)) \tag{5.10}$$

for every $t \in (t_0, t_0 + a)$. Then

$$v(t) < w(t), \quad t \in [t_0, t_0 + a)$$

Proof. Assume the conclusion is not true. Then

$$N = \{t \in [t_0, t_0 + a): w(t) \leq v(t)\} \neq \emptyset$$

Let $t_1 = \inf N$. By the continuity of w, v and the inequality $v(t_0) < w(t_0)$, we have $t_1 > t_0$.

Moreover,

$$v(t_1) = w(t_1)$$

and

$$v(t) < w(t), \quad t \in [t_0, t_1)$$

Thus, for $h < 0$ and $-h$ sufficiently small we have

$$\frac{v(t_1 + h) - v(t_1)}{h} > \frac{w(t_1 + h) - w(t_1)}{h}$$

which yields $D_-v(t_1) \geq D_-w(t_1)$. This and (5.10) imply $\gamma(t_1, v(t_1)) > \gamma(t_1, w(t_1))$; that is, a contradiction. It follows that N is the empty set.

Theorem 5.5 (Existence of maximal and minimal solutions). Let $D = [t_0, t_0 + a] \times \{u \in R; |u - u_0| \leq b\}$, where $t_0 \geq 0$, $b > 0$. Let $\gamma: D \to R$ be continuous and such that $|\gamma(t, u)| \leq K$ (K constant) for $(t, u) \in D$. Then

$$u' = \gamma(t, u), \quad u(t_0) = u_0 \tag{*}$$

has a maximal and a minimal solution on $[t_0, t_0 + \alpha]$, where $\alpha = \min\{a, b/(2K + b)\}$.

Proof. We will only be concerned with the existence of a maximal solution. Dual arguments cover the existence of a minimal solution. Let ϵ be such that $0 < \epsilon \leq b/2$ and consider the scalar problem

$$u' = \gamma(t, u) + \epsilon, \quad u(t_0) = u_0 + \epsilon \tag{5.11}_\epsilon$$

We notice that the function $\gamma(t, u) + \epsilon$ is continuous on the set D_ϵ of the points (t, u) such that $t \in [t_0, t_0 + a]$ and $|u - (u_0 + \epsilon)| \leq b/2$. We also notice that $D_\epsilon \subset D$ and that $|\gamma(t, u) + \epsilon| \leq K + (b/2)$ on D_ϵ. Consequently, the Peano theorem (Theorem 3.1) ensures the existence of a solution $u_\epsilon(t)$ of the problem $(5.11)_\epsilon$ which exists on $[t_0, t_0 + \alpha]$. The boundedness of $\gamma(t, u)$ on D implies the equicontinuity of the family of functions $u_\epsilon(t)$, $0 < \epsilon \leq b/2$, $t \in [t_0, t_0 + \alpha]$. Since these functions are also uniformly bounded, we may choose a decreasing sequence $\{\epsilon_n\}$ such that $\epsilon_n \to 0$ as $n \to \infty$ and $s(t) = \lim_{n \to \infty} u_{\epsilon_n}(t)$ exists uniformly on $[t_0, t_0 + \alpha]$ (see Theorem 2.5). Letting $n \to \infty$ in

$$u_{\epsilon_n}(t) = u_0 + \epsilon_n + \int_{t_0}^{t} [\gamma(s, u_{\epsilon_n}(s)) + \epsilon_n] ds$$

and taking into consideration the fact that $\gamma(t, u_{\epsilon_n}(t)) \to \gamma(t, s(t))$ uniformly on $[t_0, t_0 + \alpha]$, we obtain that $s(t)$ is a solution of (*) on the interval $[t_0, t_0 + \alpha]$. To show that $s(t)$ is the maximal solution of (*) on $[t_0, t_0 + \alpha]$, let $u(t)$ be any solution of (*) on $[t_0, t_0 + \alpha']$, $\alpha' \leq \alpha$. Then we have

$$u(t_0) = u_0 < u_0 + \epsilon = u_\epsilon(t_0)$$

$$u'(t) < \gamma(t, u(t)) + \epsilon$$

$$u'_\epsilon(t) = \gamma(t, u_\epsilon(t)) + \epsilon$$

for every $t \in [t_0, t_0 + \alpha']$, $\epsilon \in (0, b/2]$. Theorem 5.4 applies now to obtain $u(t) \leq u_\epsilon(t)$, $t \in [t_0, t_0 + \alpha']$. This of course implies that $u(t) \leq s(t)$ on $[t_0, t_0 + \alpha']$ and completes the proof.

The following three lemmas will be used in the proof of the "comparison theorem" (Theorem 5.10). The letter D denotes any one of the Dini derivatives.

Lemma 5.6 Let $u: [t_0, t_0 + a] \to R$, $t_0 \geq 0$, be continuous and such that $Du(t) \leq 0$ for $t \in [t_0, t_0 + a) \setminus S$, where S is a countable set. Then u is decreasing on $[t_0, t_0 + a]$.

The proof of this lemma follows easily from the proof of Theorem 34.1 in McShane [30, p. 200].

Lemma 5.7 Let $v, w: [t_0, t_0 + a] \to R$, $t_0 \geq 0$, be continuous and such that $Dv(t) \leq w(t)$ for every $t \in [t_0, t_0 + a) \setminus S$, where S is a countable set. Then $D_-v(t) \leq w(t)$ for every $t \in (t_0, t_0 + a)$.

Proof. Let

$$y(t) = v(t) - \int_{t_0}^{t} w(s) ds, \quad t \in [t_0, t_0 + \alpha]$$

Then $Dy(t) = Dv(t) - w(t) \leq 0$, $t \in [t_0, t_0 + a) \setminus S$. Consequently, Lemma 5.6 implies that $y(t)$ is decreasing on $[t_0, t_0 + a)$, which in turn yields

$$D_-y(t) = D_-v(t) - w(t) \leq 0, \quad t \in (t_0, t_0 + a)$$

(see McShane [30, p. 191]). This completes the proof of the lemma.

Remark 5.8 It is important to note now that, in view of the above lemma, the derivative D_- in Theorem 5.4 can be replaced by any other Dini derivative.

Lemma 5.9 Let $\gamma: R_+ \times R \to R$ be continuous and let $(t_0, u_0) \in R_+ \times R$ be fixed. Let $s(t)$ be the maximal solution of (*) on the interval $[t_0, t_0 + a)$, for some $a \in (0, +\infty]$, and fix $t_1 \in (t_0, t_0 + a)$. Then there exists $\epsilon_0 > 0$ such that if $0 < \epsilon < \epsilon_0$, the maximal solution $s_\epsilon(t)$ of

$$u' = \gamma(t, u) + \epsilon, \quad u(t_0) = u_0 + \epsilon \tag{5.11}_\epsilon$$

exists on $[t_0, t_1]$ and satisfies

$$\lim_{\epsilon \to 0} s_\epsilon(t) = s(t)$$

uniformly on $[t_0, t_1]$.

Proof. Let M_1 be open, bounded and $b > 0$ be such that

$$D_\epsilon^t = [t, t + b] \times \{u \in R; \; |u - (s(t) + \epsilon)| \leq b\} \in M_1$$

for every $t \in [t_0, t_1]$ and every $\epsilon \in (0, b/2]$. Suppose that $|\gamma(t, u)| \leq K$ on M_1, where K is a positive constant. Then we have

$$|\gamma(t, u) + \epsilon| \leq K + b/2$$

on D_ϵ^t for every $t \in [t_0, t_1]$, $\epsilon \in (0, b/2]$. Theorem 5.5 applied to the rectangle $D_\epsilon^{t_0}$ ensures the existence of the maximal solution $s_\epsilon(t)$ of $(5.11)_\epsilon$ on the interval $[t_0, t_0 + \alpha]$ with $\alpha = \min\{b, 2b/(2K + b)\}$. The number α does not depend on ϵ. Now we choose an integer N such that $\alpha' = (t_1 - t_0)/N \leq \alpha$, and we proceed as in Theorem 5.5 to conclude that

$$\lim_{\epsilon \to 0} s_\epsilon(t) = s(t)$$

uniformly on $[t_0, t_0 + \alpha']$. Here we use the fact that a maximal solution is unique and that for every sequence $\{\epsilon_n\}$ with $\epsilon_n > 0$ and $\lim_{n \to \infty} \epsilon_n = 0$ there exists a subsequence $\{\epsilon_n'\}$ such that $s_{\epsilon_n'}(t) \to s(t)$ as $n \to \infty$ uniformly on $[t_0, t_0 + \alpha']$. It follows that $s_\epsilon(t_0 + \alpha') \to s(t_0 + \alpha')$ as $\epsilon \to 0$. Thus, there exists a positive $\epsilon_1 \leq b/2$ such that

$$0 < \mu(\epsilon) = s_\epsilon(t_0 + \alpha') - s(t_0 + \alpha') \leq b/2, \ 0 < \epsilon \leq \epsilon_1$$

Repeating the above argument on the rectangle $D_{\mu(\epsilon)}^{t_0+\alpha'}$, $\epsilon < \epsilon_1$, we obtain that the problem

$$u' = \gamma(t, u) + \epsilon, \ u(t_0 + \alpha') = s(t_0 + \alpha') + \mu(\epsilon)$$

has its maximal solution $\bar{s}_\epsilon(t)$ existing on $[t_0 + \alpha', t_0 + 2\alpha']$. We can extend the function $s_\epsilon(t)$ on $[t_0 + \alpha', t_0 + 2\alpha']$ by defining

$$s_\epsilon(t) = \bar{s}_\epsilon(t), \ t \in [t_0 + \alpha', t_0 + 2\alpha']$$

for $\epsilon < \epsilon_1$. Obviously, this extended $s_\epsilon(t)$ is the maximal solution of $(5.11)_\epsilon$ on $[t_0, t_0 + 2\alpha']$ and converges to $s(t)$ uniformly on this interval as $\epsilon \to 0$. Similarly we work by induction to show that there is an $\epsilon_0 = \epsilon_{N-1}$ such that the maximal solution $s_\epsilon(t)$ of $(5.11)_\epsilon$ exists on $[t_0, t_0 + N\alpha'] = [t_0, t_1]$, for $0 < \epsilon < \epsilon_0$, and converges to $s(t)$ uniformly on $[t_0, t_1]$ as $\epsilon \to 0$. This completes the proof.

Theorem 5.10 (The comparison theorem). Let $\gamma: R_+ \times R \to R$ be continuous and fix $(t_0, u_0) \in R_+ \times R$, $a \in (0, +\infty]$. Let $s(t)$ be the maximal solution of (*) on the interval $[t_0, t_0 + a)$. Let $u: [t_0, t_0 + a) \to R$ be continuous and such that $u(t_0) \leq u_0$ and

$$Du(t) \leq \gamma(t, u(t)), \ t \in [t_0, t_0 + a) \setminus S \quad (5.12)$$

where S is a countable set. Then

$$u(t) \leq s(t), \ t \in [t_0, t_0 + a)$$

Proof. Let us first notice that, by Lemma 5.7, we have

$$D_-u(t) \leq \gamma(t, u(t)), \ t \in (t_0, t_0 + a) \quad (5.13)$$

If t_1 is a point in $(t_0, t_0 + a)$, then the above lemma ensures the existence of the maximal solution $s_\epsilon(t)$ of $(5.11)_\epsilon$ on $[t_0, t_1]$ for all sufficiently small $\epsilon > 0$. We also have that $s_\epsilon(t) \to s(t)$ as $\epsilon \to 0$ uniformly on $[t_0, t_1]$. Combining $(5.11)_\epsilon$, (5.13) and Theorem 5.4, we find that $u(t) \leq s_\epsilon(t)$ on $[t_0, t_1]$. This implies that $u(t) \leq s(t)$, $t \in [t_0, t_1]$. Since t_1 is arbitrary, the theorem is proved.

We are now ready to state and prove the "comparison principle" mentioned in the introduction. This principle is the source of an abundance of local and asymptotic properties of solutions of (E). Some of its applications will be given in Sections 3 and 4.

Theorem 5.11 (The comparison principle). Let V be a Lyapunov function defined on $R_+ \times R^n$ and assume that

$$V'_E(t, u) \leq \gamma(t, V(t, u)), \quad (t, u) \in R_+ \times R^n \tag{A}$$

where $\gamma: R_+ \times R \to R$ is continuous and such that the problem (*) has the maximal solution $s(t)$ on the interval $[t_0, T)$, $0 \leq t_0 < T \leq +\infty$. Let $x(t)$, $t \in [t_0, T)$ be any solution of (E) with $V(t_0, x(t_0)) \leq u_0$. Then $V(t, x(t)) \leq s(t)$ for $t \in [t_0, T)$.

Proof. Let $\lambda(t) = V(t, x(t))$, $t \in [t_0, T)$. Then $\lambda(t_0) \leq u_0$ and, by Lemma 5.2,

$$D^+\lambda(t) \leq \gamma(t, \lambda(t)), \quad t \in [t_0, T)$$

Consequently, Theorem 5.10 implies $V(t, x(t)) \leq s(t)$, $t \in [t_0, T)$.

3. Existence on R_+

In this section we employ the comparison principle in order to establish an existence theorem on R_+ for the system (E). The following lemma is an easy consequence of the proofs of Theorems 3.6 and 3.7.

Lemma 5.12 Let $x(t)$, $t \in [t_0, t_1)$, $0 \leq t_0 < t_1 < +\infty$, be a solution of the system (E). Then $x(t)$ is extendable to the point $t = t_1$ if and only if it is bounded on $[t_0, t_1)$.

Definition 5.13 Let $x(t)$, $t \in [t_0, T)$, $0 \leq t_0 < T \leq +\infty$, be a solution of the system (E). Then $x(t)$ is said to be "noncontinuable" or "nonextendable" to the right if T equals $+\infty$ or if $x(t)$ cannot be continued to the point T.

Theorem 5.14 says that every extendable to the right solution of the system (E) is part of a noncontinuable to the right solution of the same system.

Theorem 5.14 Let $x(t)$, $t \in [t_0, t_1)$, $t_1 > t_0 \geq 0$, be an extendable to the right solution of the system (E). Then there exists a noncontinuable to the right solution of (E) which extends $x(t)$; that is, a solution $y(t)$, $t \in [t_0, t_2)$ such that $t_2 > t_1$, $y(t) = x(t)$, $t \in [t_0, t_1)$, and $y(t)$ is noncontinuable to the right. Here t_2 may equal $+\infty$.

Proof. It suffices to assume that $t_0 > 0$. Let $Q = (0, \infty) \times R^n$ and, for $m = 1, 2, \ldots$, let $Q_m = \{(t, u) \in Q; t^2 + \|u\|^2 \leq m, t \geq 1/m\}$. Then $Q_m \subset Q_{m+1}$ and $\cup Q_m = Q$. Furthermore, each Q_m is a compact subset of Q. By Lemma 5.12, a solution $y(t)$, $t \in [t_0, T)$, is noncontinuable to the right, if its graph $\{(t, y(t)); t \in [t_0, T)\}$ intersects all the sets Q_m. We are going to construct such a solution $y(t)$ which extends $x(t)$. Since $x(t)$ is continuable to the right, we may consider it defined and continuous on the interval $[t_0, t_1]$. Now since the graph $G = \{(t, x(t)); t \in [t_0, t_1]\}$ is compact, there exists an integer m_1 such that $G \in Q_{m_1}$. If the number $\alpha > 0$ is sufficiently small, then for every $(a, u) \in Q_{m_1}$ the set

$$M_{a,u} = \{(t, x) \in R^{n+1}; \ |t - a| \leq \alpha, \ \|x - u\| \leq \alpha\}$$

is contained in the set Q_{m_1+1}. Let $\|f(t, x)\| \leq K$ on the set Q_{m_1+1}, where K is a positive constant. By Peano's theorem (Theorem 3.1), for every point $(a, u) \in Q_{m_1}$ there exists a solution $x(t)$ of the system (E), such that $x(a) = u$, defined on the interval $[a, a + \beta]$ with $\beta = \min\{\alpha, \alpha K^{-1}\}$. This number β does not depend on the particular point $(a, u) \in Q_{m_1}$. Consequently, since $(t_1, x(t_1)) \in Q_{m_1}$, there exists a solution $x_1(t)$ of (E) which continues $x(t)$ to the point $t_1 + \beta$. Repeating this process, we shall eventually have a solution $x_q(t)$ (q a positive integer) of the system (E), which continues $x(t)$ to the point $t_0 + q\beta$ and has a graph in the set Q_{m_1+1}, but not entirely inside the set Q_{m_1}. In this set Q_{m_1+1} we repeat the continuation process as in the set Q_{m_1}. Thus, we eventually obtain a solution $y(t)$ which intersects all the sets Q_m, $m \geq m_1$, for some m_1, and is a noncontinuable extension of the solution $x(t)$.

The theorem below shows that a noncontinuable to the right solution of (E) actually "blows up" at the right endpoint T of the interval of its existence, provided this point T is finite.

Theorem 5.15 Let $x(t)$, $t \in [t_0, T)$ $(0 \leq t_0 < T < +\infty)$, be a noncontinuable to the right solution of (E). Then $\lim_{t \to T^-} \|x(t)\| = +\infty$.

Proof. Assume that our assertion is false. Then there exists an increasing sequence $\{t_m\}_{m=1}^{\infty}$ such that $t_0 \leq t_m < T$, $\lim_{m \to \infty} t_m = T$ and $\lim_{m \to \infty} \|x(t_m)\| = L < +\infty$. Since $\{x(t_m)\}$, $m = 1, 2, \ldots$, is bounded, there exists a subsequence $\{x(t'_m)\}$ such that $x(t'_m) \to y$ as $m \to \infty$, with $\|y\| = L$ and t'_m increasing. Let M be a compact subset of $R_+ \times R^n$ such that the point (T, y) is an interior point of M. Then we may assume that $(t'_m, x(t'_m)) \in M$ for $m = 1, 2, \ldots$. We shall show that, for infinitely many m, there exists $\bar{t}_m$ such that

$$t'_m < \bar{t}_m < t'_{m+1}, \quad (\bar{t}_m, x(\bar{t}_m)) \in \partial M \tag{5.14}$$

where ∂M denotes the boundary of M. In fact, if this were not the case, then there would be $\epsilon \in (0, T)$ such that $(t, x(t))$ belongs to the interior of M for all t with $T - \epsilon < t < T$. But then Lemma 5.12 implies the extendability of $x(t)$ beyond T, which is a contradiction. Let (5.14) hold for a subsequence $\{m'\}$ of the positive integers; that is,

$$t'_m < \bar{t}_m' < t'_{m'+1}, \quad (\bar{t}_m', x(\bar{t}_m')) \in \partial M$$

Then we have $\lim_{m \to \infty} (\bar{t}_m', x(\bar{t}_m')) = (T, y)$. This is a consequence of the fact that $\bar{t}_m' \to T$ as $m' \to \infty$ and the inequality $\|x(t'_m) - x(\bar{t}_m')\| \le \mu(\bar{t}_m' - t'_m)$, where μ is a bound for F on M. However, ∂M is a closed set. Thus, the point $(T, y) \in \partial M$, a contradiction to our assumption. This completes the proof.

It should be noted that a local maximal solution of (*) can always be extended to a noncontinuable to the right maximal solution. This is a consequence of Theorems 5.5 and 5.14.

We are now ready for the main result of this section, which ensures the existence on R_+ of the solution of a system, by using this property for an associated scalar equation.

Theorem 5.16 (Comparison principle and existence on R_+). Let $V: R_+ \times R^n \to R$ be a Lyapunov function satisfying

$$V_E'(t, u) \le \gamma(t, V(t, u)), \quad (t, u) \in R_+ \times R^n \tag{5.15}$$

and $V(t, u) \to +\infty$ as $\|u\| \to +\infty$ uniformly w.r.t. t lying in any compact set. Here $\gamma: R_+ \times R \to R$ is continuous and such that for every $(t_0, u_0) \in R_+ \times R$ the problem (*) has a maximal solution defined on $[t_0, +\infty)$. Then every solution of (E) is extendable to $+\infty$.

Proof. Let $[t_0, T)$ be the maximal interval of existence of a solution $x(t)$ of (E) and assume that $T < +\infty$. Let $y(t)$ be the maximal solution of (*) with $y(t_0) = V(t_0, x(t_0))$. Then Theorem (5.11) ensures that

$$V(t, x(t)) \le y(t), \quad t \in [t_0, T) \tag{5.16}$$

On the other hand, since $x(t)$ is a noncontinuable to the right solution, we must have

$$\lim_{t \to T^-} \|x(t)\| = +\infty$$

by Theorem 5.15. This implies that $V(t, x(t))$ converges to $+\infty$ as $t \to T^-$, but (5.16) implies that $\limsup V(t, x(t)) \le y(T)$ as $t \to T^-$. Thus, $T = +\infty$.

It is easy to see that (5.15) is not needed for all $(t, u) \in R_+ \times R^n$. It can be assumed instead that it holds for all $u \in R^n$ such that $\|u\| > \alpha$, where α is a positive constant. Having this in mind, we establish the important Corollary 5.17.

Corollary 5.17 Assume that there exists $\alpha > 0$ such that

$$\|F(t, u)\| \leq \gamma(t, \|u\|), \quad t \in R_+, \|u\| > \alpha$$

where $\gamma: R_+ \times R_+ \to R_+$ is such that for every $(t_0, u_0) \in R_+ \times R_+$ the problem (*) has a maximal solution defined on $[t_0, \infty)$. Then every solution of (E) is extendable to $+\infty$.

Proof. Here it suffices to take $V(t, u) = \|u\|$. In fact, from (4.19) we obtain

$$V'_E(t, x(t)) = \lim_{h \to 0^+} \frac{\|x(t) + hF(t, x(t))\| - \|x(t)\|}{h}$$

$$\leq \|F(t, x(t))\| \leq \gamma(t, \|x(t)\|) = \gamma(t, V(t, x(t)))$$

provided that $\|x(t)\| > \alpha$. Now let $x(t)$, $t \in [t_0, T)$, be a noncontinuable to the right solution of (E) such that $T < +\infty$. Then, for t sufficiently close to T from the left, we have $\|x(t)\| > \alpha$, and the rest of the proof follows as in Theorem (5.16).

4. The Comparison Principle and Stability

In this section we establish the stability of a system of the form (E) assuming the stability of an associated scalar equation of the form (*).

Definition 5.18 Let the function $\lambda: R_+ \to R_+$ be strictly increasing, continuous, and such that $\lambda(0) = 0$. Then λ is called a "Q-function." The set of all Q-functions will be be denoted by QF.

Definition 5.19 A Lyapunov function $V: R_+ \times R^n \to R$ with $V(t, 0) \equiv 0$ is said to be "Q-positive" if there exists $\lambda \in QF$ such that

$$V(t, u) \geq \lambda(\|u\|), \quad (t, u) \in R_+ \times R^n$$

Definition 5.20 A Lyapunov function $V: R_+ \times R^n \to R$ is said to be "Q-bounded" if there exists $\lambda \in QF$ such that

$$V(t, u) \leq \lambda(\|u\|), \quad (t, u) \in R_+ \times R^n$$

We are now ready for our first stability results via the comparison method.

Theorem 5.21 Let $F(t, 0) \equiv 0$ and $V: R_+ \times R^n \to R$ be a Lyapunov function with the property

$$V'_E(t, u) \leq \gamma(t, V(t, u)), \quad (t, u) \in R_+ \times R^n$$

where $\gamma: R_+ \times R \to R$ is continuous and $\gamma(t, 0) \equiv 0$. If the zero solution of

$$u' = \gamma(t, u) \qquad (5.17)$$

is stable (asymptotically stable) and V is Q-positive, then the zero solution of (E) is stable (asymptotically stable).

Proof. Assume the stability of the zero solution of (5.17). The Q-positiveness of V implies the existence of $\lambda \in QF$ such that $V(t, u) \geq \lambda(\|u\|)$ for all $(t, u) \in R_+ \times R^n$. Given $\epsilon > 0$ there exists $\eta(\epsilon) > 0$ such that $|y(t)| < \lambda(\epsilon)$ for $t \geq 0$, where $y(t)$ is any solution of (5.17) with $|y(0)| < \eta(\epsilon)$. This property follows from the stability assumption on (5.17). On the other hand, since V is continuous and $V(t, 0) \equiv 0$, there exists $\delta(\epsilon) > 0$ such that

$$V(0, x_0) < \eta(\epsilon) \text{ whenever } \|x_0\| < \delta(\epsilon)$$

Now fix x_0 with $\|x_0\| < \delta(\epsilon)$, let $x(t)$ be a solution of (E) with $x(0) = x_0$, and let $u(t)$ be the maximal solution of (5.17) with the property $u(0) = V(0, x_0)$.

Then, by Theorem 5.11, we have

$$\lambda(\|x(t)\|) \leq V(t, x(t)) \leq u(t) < \lambda(\epsilon), \quad t \geq 0$$

which, along with the fact that λ is strictly increasing, implies that

$$\|x(t)\| < \epsilon, \quad t \geq 0$$

This proves the stability of (E).

If we assume the asymptotic stability of the zero solution of (5.17), then the asymptotic stability of the zero solution of (E) follows from

$$\lambda(\|x(t)\|) \leq u(t), \quad t \geq 0$$

which holds for $\|x_0\|$ sufficiently small. Since $\lim_{t \to \infty} u(t) = 0$, we also have

$$\lim_{t \to \infty} \|x(t)\| \leq \lim_{t \to \infty} \lambda^{-1}(u(t)) = 0$$

This completes the proof.

The cases of uniform asymptotic stability are covered by Theorem 5.22.

Theorem 5.22 Let $F(t, 0) \equiv 0$ and $V: R_+ \times R^n \to R$ be a Lyapunov function with the property:

$$V'_E(t, u) \leq \gamma(t, V(t, u)), \quad (t, u) \in R_+ \times R^n$$

where $\gamma: R_+ \times R \to R$ is continuous and $\gamma(t, 0) \equiv 0$. Assume that V is Q-positive and Q-bounded. Then if the zero solution of (5.17) is uniformly (uniformly asymptotically) stable, the same fact is true for (E).

Proof. Let $\lambda, \mu \in QF$ be such that

$$\lambda(\|u\|) \leq V(t, u) \leq \mu(\|u\|), \quad (t, u) \in R_+ \times R^n \tag{5.18}$$

and let (5.17) have its zero solution uniformly stable. Then given $\epsilon > 0$, $t_0 \geq 0$, there exists $\eta(\epsilon) > 0$, independent of t_0, such that $|y(t)| < \lambda(\epsilon)$ for all $t \geq t_0$, where $y(t)$ is any solution of (5.17) with $|y(t_0)| < \eta(\epsilon)$. Now let $\delta(\epsilon) > 0$ satisfy $\delta(\epsilon) < \mu^{-1}(\eta(\epsilon))$. Then if $x_0 \in R^n$ is given with $\|x_0\| < \delta(\epsilon)$, we have

$$V(t, x_0) \leq \mu(\|x_0\|) < \eta(\epsilon)$$

for every $t \geq t_0$. The proof now follows the steps of the proof of Theorem (5.21) by arguing at $t = t_0$ instead of $t = 0$. Now let (5.17) have its zero solution uniformly asymptotically stable, and let λ, u be as in (5.18). Then some $\eta_0 > 0$ has the following property: given $\epsilon > 0$, there exists $T(\epsilon) > 0$ such that every solution $y(t)$ of (5.17) with $|y(t_0)| < \eta_0$, for some $t_0 \geq 0$, satisfies

$$|y(t)| < \lambda(\epsilon), \quad t \geq t_0 + T(\epsilon) \tag{5.19}$$

Let $\delta_0 > 0$ be such that $\mu(\delta_0) < \eta_0$. Then if x_0 is a vector in R^n with $\|x_0\| < \delta_0$, we have

$$V(t, x_0) \leq \mu(\|x_0\|) < \mu(\delta_0) < \eta_0$$

for every $t \geq 0$. From this point on, the proof follows again as in Theorem 5.21 by arguing at the point $t = t_0 + T(\epsilon)$ instead of $t = 0$. This completes the proof.

EXERCISES

5.1. Consider (E) with $F(t, 0) \equiv 0$. Let $V: R_+ \times R^n \to R_+$ satisfy

$$a\|u\| \leq V(t, u) \leq b\|u\|, \quad V'_E(t, u) \leq -cV(t, u)$$

for every $(t, u) \in R_+ \times R^n$, where a, b, c are positive constants. Show that

$$\|x(t)\| \le \|x_0\| e^{-c(t-t_0)}, \quad t \ge t_0 \ge 0$$

where $x(t)$ is any solution of (E) with $x(t_0) = x_0$. This phenomenon is also called "exponential asymptotic stability" of the zero solution of (E).

5.2. Consider (E) with $F(t, 0) \equiv 0$. Let $V: R_+ \times R^n \to R$ be a Q-positive Lyapunov function satisfying
$$V'_E(t, u) \le - p(t)q(V(t, u))$$
for all $(t, u) \in R_+ \times R^n$, where $p: R_+ \to R_+$ is continuous and satisfies
$$\int_0^\infty p(t)dt = +\infty$$
Provide conditions on $q: R \to R$ so that the zero solution of (E) is asymptotically stable.

5.3. Consider the system
$$\begin{aligned} x'_1 &= x_2 \\ x'_2 &= -x_1 - g(x_2) \end{aligned} \quad (A)$$
where $g: R \to R$ is continuous with $g(0) = 0$. Provide conditions on g that ensure the stability of the zero solution of (A). [Hint. Use the Lyapunov function $V(t, u) \equiv \|u\|^2$.]

5.4. Consider the system
$$\begin{aligned} x'_1 &= x_2 \\ x'_2 &= -p(x_1, x_2) \end{aligned} \quad (B)$$
where $p: R^2 \to R$ is continuous and such that $\operatorname{sgn} p(u_1, 0) = \operatorname{sgn} u_1$ for any $u_1 \in R$,
$$u_2[p(u_1, 0) - p(u_1, u_2)] \le 0, \quad (u_1, u_2) \in R^n$$
and
$$\lim_{|v| \to \infty} \int_0^v p(s, 0)ds = +\infty$$
Show that all solutions of (B) are bounded. [Hint. Consider the Lyapunov function
$$V(t, u) = u_2^2 + 2 \int_0^{u_1} p(s, 0)ds$$

where $u = (u_1, u_2) \in R^n$.] (Here $\operatorname{sgn} u = u/|u|$ for $u \neq 0$ and $\operatorname{sgn} u = 0$ for $u = 0$.)

5.5. Study the stability properties of the **linear** system

$$\begin{bmatrix} x_1 \\ x_2 \end{bmatrix}' = \begin{bmatrix} b(t) & a(t) \\ -a(t) & b(t) \end{bmatrix} \begin{bmatrix} x_1 \\ x_2 \end{bmatrix}$$

with $a, b: R_+ \to R$ continuous, by imposing general conditions on the functions a, b. Use the function $V(u) \equiv \|u\|^2$.

5.6. In Theorem 5.11 replace (A) by

$$V_E'(t, u) + \lambda(\|u\|) \le \gamma(t, V(t, u))$$

where $\lambda \in QF$ and γ is increasing in its second variable. Show that the conclusion of that theorem holds with

$$V(t, x(t)) + \int_{t_0}^t \lambda(\|x(s)\|)ds$$

instead of $V(t, x(t))$.

5.7. Let F in (E) be such that $\|F(t, u)\| \le p(t)q(\|u\|)$, where $p: R_+ \to R_+$, $q: R_+ \to R_+$ are continuous with $q(0) = 0$ and $q(s) > 0$ for $s > 0$. Furthermore, let

$$\int_v^\infty \frac{ds}{q(s)} = +\infty$$

for some $v > 0$. Using

$$u' = p(t)q(u), \quad u(t_0) = u_0 > 0$$

show that for every $(t_0, x_0) \in R_+ \times R^n$ there exists a solution $x(t)$, $t \in R_+$, of (E) with $x(t_0) = x_0$.

5.8. (Integral inequalities). Let $K: J \times J \times R \to R$, $J = [t_0, \infty)$, be continuous and increasing in its third variable. Assume further that for three continuous functions $u, v, f: J \to R$ we have

$$u(t) < f(t) + \int_{t_0}^t K(t, s, u(s))ds$$

$$v(t) \ge f(t) + \int_{t_0}^t K(t, s, v(s))ds$$

for every $t \ge t_0$ and $u(t_0) < v(t_0)$. Show that $u(t) < v(t)$, $t \ge t_0$. [Hint.

Assume that there is $t_1 > t_0$ with $u(t_1) = v(t_1)$ and $u(t) < v(t)$, $t \in [t_0, t_1)$. Prove that this assumption leads to the contradiction $u(t_1) < v(t_1)$.]

5.9. (Local solutions to scalar integral equations). Consider the equation

$$x(t) = f(t) + \int_{t_0}^{t} K(t, s, x(s))ds \qquad (I_1)$$

where $f: [t_0, t_0 + a] \to R$ is continuous. Furthermore, let $K: D \to R$ be continuous, where

$$D = [t_0, t_0 + a] \times [t_0, t_0 + a] \times \{u \in R; |u - f(t)| \le b, \text{ for some } t \in [t_0, t_0 + a]\}$$

with b a positive constant. Show that (I_1) has a solution $x(t)$, $t \in [t_0, t_0 + \alpha]$, where $\alpha = \min\{a, b/K_0\}$, $K_0 = \sup\{|K(t, s, u)|; (t, s, u) \in D\}$. [Hint. Consider the sequence $\{x_m(t)\}_{m=1}^{\infty}$ with $x_1(t) = f(t)$, $t \in [t_0, t_0 + \alpha]$, and

$$x_m(t) = f(t), \quad t \in [t_0, t_0 + (\alpha/m)]$$

$$x_m(t) = f(t) + \int_{t_0}^{t - (\alpha/m)} K(t, s, x_m(s))ds, \quad t \in [t_0 + (\alpha/m), t_0 + \alpha]$$

for every $m = 2, 3, \ldots$. The first equation defines $x_m(t)$ on $[t_0, t_0 + (\alpha/m)]$. From the second equation we get

$$x_m(t) = f(t) + \int_{t_0}^{t-(\alpha/m)} K(t, s, f(s))ds, \quad t \in [t_0 + (\alpha/m), t_0 + (2\alpha/m)]$$

In the next step, $x_m(t)$ is defined on $[t_0 + (2\alpha/m), t_0 + (3\alpha/m)]$ and so on. Thus, the second equation of definition of $x_m(t)$ defines $x_m(t)$ "piecewise" on the intervals $[t_0 + k\alpha/m, t_0 + (k+1)\alpha/m]$, $k = 1, 2, \ldots$, $m = 1$. The resulting function is continuous on $[t_0, t_0 + \alpha]$. Show that $|x_m(t) - f(t)| \le b$ and use Theorem 2.5 to show the existence of a subsequence of $\{x_m(t)\}$ converging to the desired solution.]

5.10. (Maximal solutions of intergral equations). Let the assumptions of Exercise 5.9 be satisfied with the last factor of D replaced by R. Assume further that K is increasing in its last variable. Then (I_1) has a maximal solution on some interval $[t_0, t_0 + \alpha]$, $\alpha > 0$. [Hint. Model your proof after that of Theorem 5.5. Consider the integral equation

$$x(t) = f(t) + \epsilon + \int_{t_0}^{t} K(t, s, x(s))ds$$

Apply the result of Exercise 5.9 to obtain a solution $x_\epsilon(t)$ on $[t_0, t_0 + \alpha]$.]

5.11. Let $F: R_+ \times R^n \to R^n$, $V: R^n \to R_+$ be continuous and such that $F(t, 0) \equiv 0$, $V(0) = 0$ and $V(u) > 0$ for $u \neq 0$. Assume further that V is continuously differentiable on the ball $S_\alpha = \{u \in R^n; \|u\| < \alpha\}$ and satisfies

$$\langle \nabla V(u), F(t, u) \rangle \leq 0 \tag{D}$$

for all $t \in R_+$, $u \in S_\alpha$. Prove that the zero solution of

$$x' = F(t, x) \tag{E}$$

is stable by following these steps: letting $\epsilon \in (0, \alpha)$, show that V attains its minimum on $S = \{u \in R^n; V(u) = \epsilon\}$. Then show that there is $\delta \in (0, \epsilon)$ such that $\|x(0)\| \leq \delta$ implies that $x(t)$ cannot reach the surface S. (Naturally, this problem can be solved by considering the scalar equation $u' = 0$.)

5.12. Let the assumptions of Exercise 5.11 be satisfied with $F(t, u) \equiv F(u)$. Assume further that

$$\langle \nabla V(u), F(u) \rangle < 0$$

for every $t \in R_+$, $u \in S_\beta$ such that $u \neq 0$. Here $\beta \in (0, \alpha]$. Show that the zero solution of (E) is asymptotically stable. [Hint. Use the fact that the function $\langle \nabla V(u), F(u) \rangle$ attains its maximum on every set $S_{\lambda, \mu} = \{u \in R^n; \lambda \leq \|u\| \leq \mu \leq \beta\}$, where $\lambda > 0$, μ are constants.]

5.13. Consider Van der Pol's equation

$$x'' = k(1 - x^2)x' + x = 0 \tag{G}$$

where k is a constant.
 (1) Determine k so that the zero solution of (G) (written as a system) is stable.
 (2) Extend your result to Lienard's equation

$$x'' + f(x)x' + g(x) = 0 \tag{L}$$

with $f: R \to R$, $g: R \to R$ continuous and $g(0) = 0$.
 (3) Impose further conditions on f, g that guarantee the asymptotic stability of the zero solution of (L). [Hint. Let

$$x_1' = x_2$$

$$x_2' = -f(x_1)x_2 - g(x_1)$$

and use the Lyapunov function

$$V(x_1, x_2) = (1/2)x_2^2 + \int_0^{x_1} g(u)du$$

in connection with Exercises 5.11 and 5.12.]

5.14. (Instability). Let $V: R^n \to R_+$, $F: R^n \to R^n$ be continuous with $F(0) = 0$, $V(0) = 0$ and $V(u) > 0$ for $u \neq 0$. Furthermore, assume that V is continuously differentiable on $S_\alpha = \{u \in R^n; \|u\| < \alpha\}$, for some $\alpha \in (0, \infty)$, and such that

$$<\nabla V(u), F(u)> \; > 0$$

for all $u \in S_\alpha$ with $u \neq 0$. Show that the zero solution of (E) is unstable. [Hint. let $\beta \in (0, \alpha)$, and let M be the maximum of V on $\bar{S}_\beta = \{u \in R^n; \|u\| \leq \beta\}$. Let $\delta \in (0, \beta)$ and choose $c \in S_\delta$ $(c \neq 0)$, $\mu \in (0, \|c\|)$ such that $V(u) < V(c)$ for $\|u\| \leq \mu$. Let m be the minimum of $<\nabla V(u), F(u)>$ on $\{u \in R^n; \mu \leq \|u\| \leq \alpha\}$. Show that if the solution $x(t)$ satisfies $x(0) = c$, then

$$V(x(t)) \geq V(c) + tm$$

as long as $\|x(t)\| \leq \beta$. This implies that $x(t)$ goes through the surface $S = \{u \in R^n; \|u\| = \beta\}$ at some time t_0.]

5.15. Let $F: R \times R^n \to R^n$, $V: R^n \to R_+$ be continuous and such that $F(t, u)$ is T-periodic in t, $F(t, 0) \equiv 0$, $V(0) = 0$ and $V(u) > 0$ for $u \neq 0$. Assume that V is continuously differentiable on R^n and satisfies

$$<\nabla V(u), F(t, u)> \; \leq 0$$

for all $t \in [0, T]$, $u \in R^n$. Assume further that, for any constant $c > 0$, there is no solution $x(t)$ of (E) such that $V(x(t)) = c$ for all $t \in [0, T]$. Show that (E) has no T-periodic solutions $x(t) \not\equiv 0$.

5.16. (Uniqueness via Lyapunov functions). Let $V: R^n \to R_+$, $F: R_+ \times R^n \to R^n$ be continuous and such that $V(0) = 0$, $V(u) > 0$ for $u \neq 0$, and $F(t, u) \equiv 0$. Furthermore, let V be continuously differentiable on R^n and such that

$$<\nabla V(u - v), F(t, u) - F(t, v)> \; \leq 0$$

for any $t \in R_+$, $(u, v) \in R^n \times R^n$. Then if $x(t)$, $y(t)$, $t \in [t_0, \infty)$, $t_0 \geq 0$, are two solutions of (E) such that $x(t_0) = y(t_0)$, we have $x(t) = y(t)$, $t \in [t_0, \infty)$. Show this, and then consider an improvement of this result by replacing zero in the above inequality by

$\gamma(t, V(u - v))$, for some suitable scalar function γ.

5.17. Extend the results of Execises 5.12, 5.14 to (non-autonomous) systems (E) with F depending also on the variable t. Furthermore, examine the case of Lyapunov functions V depending also on the variable t.

5.18. Prove Lemma 5.12.

CHAPTER SIX

BOUNDARY VALUE PROBLEMS ON FINITE AND INFINITE INTERVALS

Let J be a subinterval of R, and let $F: J \times R^n \to R^n$ be continuous. Let $A(J)$ be a class of continuous R^n-valued functions defined on J. Then the system

$$x' = F(t, x) \tag{E}$$

along with the condition $x \in A(J)$ is a "boundary value problem" on the interval J. Of course, the condition $x \in A(J)$ is too general and contains the initial value problem on $J = [t_0, T]$, that is, $A(J)$ can be the class $\{x \in C_n[t_0, T]; x(t_0) = x_0\}$. A boundary value problem (b.v.p.) usually concerns itself with "boundary conditions" $x \in A(J)$ that involve values of x at more than one point of the interval J. One of the most important b.v.p.'s in the theory of ordinary differential equations is the problem concerning the existence of a periodic solution. This problem consists of (E) and the condition

$$x(t + T) = x(t) \text{ for every } t \in R \tag{B_1}$$

and some fixed number $T > 0$. Here we usually assume that F is T-period in its first variable t which ranges over R. Another more general b.v.p. consists of (E) and the condition

$$Mx(0) - Nx(T) = 0 \tag{B_2}$$

where M, N are known constant $n \times n$ matrices. Naturally, the points $0, T$

can be replaced here by any other points a, b in R with $b > a$. In case $M = N = I$ the condition (B$_2$) coincides with (B$_1$). An even more general problem than ((E), (B$_2$)) is the problem consisting of (E) and the condition

$$Ux = r \tag{B$_3$}$$

Here r is a fixed vector of R^n and the operator U is defined on a subspace of $C_n[0, T]$ and has values in R^n. Such an operator could be given, for example, by

$$Ux = \int_0^T V(s)x(s)ds$$

where $V: [0, T] \to M_n$ is continuous. If the operator U in (B$_3$) is defined on a class $A(J)$, with J an infinite interval, then we have an example of a b.v.p. on an infinite interval. For example,

$$Ux = Mx(0) - Nx(\infty)$$

where $x(\infty)$ denotes the limit of $x(t)$ as $t \to \infty$ and M, N are as in (B$_2$). Here we may consider $A(J) = \{u \in C_n^1, Uu = r\}$.

In this chapter we study b.v.p.'s on finite as well as infinite intervals. We begin with the case of linear systems and continue with perturbed systems of the type studied in Chapter Four. All the boundary value problems on finite (closed) intervals in this chapter are studied on $[0, T]$. Extensions to arbitrary finite (closed) intervals are trivial.

1. Linear Systems; Finite Intervals

In this section we study b.v.p.'s for linear systems of the forms

$$x' = A(t)x, \tag{S}$$

$$x' = A(t)x + f(t) \tag{S$_f$}$$

where $A: J \to M_n$, $f: J \to R^n$ are assumed to be continuous on the interval $J = [0, T]$. Let $X(t)$ be the fundamental matrix of (S) with the property $X(0) = I$. Then the general solution of (S) is $X(t)x_0$, where x_0 is an arbitrary vector in R^n. Let $U: C_n[0, T] \to R^n$ be a bounded linear operator. Then $U(X(\cdot)x_0) = \tilde{X}x_0$ for every $x_0 \in R^n$, where $\tilde{X}$ is the matrix whose columns are the values of U on the corresponding columns of $X(t)$. It is easy to prove that this equation holds. It is obvious that the homogeneous problem (S) with the homogeneous conditions

$$Ux = 0 \tag{B$_4$}$$

is satisfied only by the zero solution if and only if $x_0 = 0$ is the only solution of $\widetilde{X}x_0 = 0$; that is, if and only if $\widetilde{X}$ is nonsingular. Now let us look at the problem $((S_f), (B_3))$. The general solution of (S_f) is given by

$$x(t) = X(t)x_0 + p(t, f), \quad t \in [0, T] \tag{6.1}$$

where

$$p(t, f) = X(t) \int_0^t X^{-1}(s) f(s) ds$$

The solution $x(t)$, with $x(0) = x_0$, will satisfy (B_3) if and only if

$$Ux = r = \widetilde{X}x_0 + Up(\cdot, f)$$

This equation in x_0 has a unique solution in R^n for some $f \in C_n[0, T]$, $r \in R^n$, if and only if $\widetilde{X}$ is nonsingular, and this solution has initial condition x_0 given by

$$x_0 = \widetilde{X}^{-1}[r - Up(\cdot, f)] \tag{6.2}$$

Consequently, we have shown the following theorem.

Theorem 6.1 Consider the problem $((S_f), (B_3))$, where $U: C_n[0, T] \to R^n$ is a bounded linear operator. Then the following statements are equivalent:

(i) the problem $((S), (B_3))$ is satisfied only by the zero solution of (S);
(ii) the problem $((S_f), (B_3))$ has a unique solution for every $(r, f) \in R^n \times C_n[0, T]$;
(iii) the problem $((S_f), (B_3))$ has a unique solution for some $(r, f) \in R^n \times C_n[0, T]$;
(iv) $\widetilde{X}$ is nonsingular.

This theorem holds true if $C_n[0, T]$ is replaced by $C_n(R_+)$ or $C_n^1(R_+)$, provided, of course, that the matrices $\widetilde{X}$ and $Up(\cdot, f)$ are well defined.

2. Periodic Solutions of Linear Systems

We now show that the dimension of the vector space of periodic solutions $(x(0) = x(T))$ of (S) is the same as the dimension of the corresponding space of the "adjoint system:"

$$y' = -yA(t) \tag{S_a}$$

where $y = [y_1, y_2, \ldots, y_n]$. Whenever we are dealing with T-periodicity, $F(t, x)$ in (E) and $A(t)$, $f(t)$ in (S_f) will be assumed to be T-periodic in t. The reader should not confuse the period T with the letter T indicating transpose. For a horizontal vector Y, the symbol y^T denotes the corresponding vertical vector.

102 CHAPTER SIX

Theorem 6.2 Let m be the number of linearly independent T-periodic solutions of the system (S). Then m is also the number of linearly independent T-periodic solutions of (S$_a$).

Proof. The system (S) has a T-periodic solution with initial value x_0 if we have $x(T)x_0 = x_0$. From Exercise 3.4 we obtain that $X^{-1}(t)$ is the fundamental matrix of the system (S$_a$) with $X^{-1}(0) = I$. Consequently, the general solution of (S$_a$) is $y(t) = y_0 X^{-1}(t)$. The system (S$_a$) will have a T-periodic solution with initial value y_0 if we have $y_0 = y_0 X^{-1}(T)$ or $y_0 X(T) = y_0$. Transposing this equation we obtain $X^T(T)y_0^T = y_0^T$. However, the matrices $X(T) - I$, $X^T(T) - I$ have the same rank. It follows that the equations

$$[X(T) - I]x_0 = 0, \quad [X^T(T) - I]y_0^T = 0$$

have the same number of linearly independent solutions. This means that the systems (S), (S$_a$) have the same number of linearly independent T-periodic solutions.

Theorem 6.3 is called the "Fredholm alternative" and provides a necessary and sufficient condition for the existence of T-periodic solutions of the system (S$_f$).

Theorem 6.3 A necessary and sufficient condition for the existence of a T-periodic solution of the system (S$_f$) is that f be orthogonal (in the integral sense) to all periodic solutions of (S$_a$), that is,

$$\int_0^T y_j(t)f(t)dt = 0, \quad j = 1, 2, \ldots, m \qquad (6.3)$$

where $\{y_j\}$, $j = 1, 2, \ldots, m$, is a basis for the vector space of T-periodic solutions of the system (S$_a$).

Proof. The general solution of (S$_f$) is given by

$$x(t) = X(t)x_0 + \int_0^t X(t)X^{-1}(s)f(s)ds$$

for every $t \in [0, T]$, $x_0 \in R^n$. Thus,

$$x(T) = X(T)x_0 + X(T)\int_0^T X^{-1}(s)f(s)ds$$

Hence, $x(t)$ will be periodic if and only if

$$[I - X(T)]x_0 = X(T)\int_0^T X^{-1}(s)f(s)ds \qquad (6.4)$$

has solutions x_0. Assume that (S$_f$) does have a T-periodic solution $x(t)$ with initial condition $x(0) = x_0$. Let $y(t)$ be a T-periodic solution of (S_a) with $y(0) = y_0$. Then we have (cf. proof of Theorem 6.2) $y_0 = y_0 X(T)$ or

$$y_0[I - X(T)] = 0 \qquad (6.5)$$

It follows that

$$y_0[I - X(T)]x_0 = 0 \qquad (6.6)$$

Equations (6.4) and (6.6) yield

$$y_0 X(T) \int_0^T X^{-1}(s)f(s)ds = \int_0^T y_0 X(T) X^{-1}(s)f(s)ds = 0 \qquad (6.7)$$

However, $y_0 X(T) = y_0$ and $y_0 X^{-1}(t) = y(t)$ (cf. proof of Theorem 6.2).

Hence,

$$\int_0^T y(s)f(s)ds = 0 \qquad (6.8)$$

which shows that (6.3) is necessary.

Before we show that (6.3) is also sufficient, let us first recall that a system

$$xC = D \quad (Cx = D)$$

with $C \in M_n$, x, D row R^n-vectors (column R^n-vectors), has at least one solution if and only if the rank of C equals the rank of the augmented matrix $[C \vdots D]^{(*)}$. Now assume that (6.8) is true for every T-periodic solution $y(t)$ of (S_a). Then we have

$$y_0 X(T) \int_0^T X^{-1}(s)f(s)ds = 0$$

for every solution y_0 of the system

$$y_0 = y_0 X(T)$$

This implies that the dimension of the solution space of (6.5) is the same as the dimension of the solution space of the system

$$y_0[I - X(T)] = 0, \quad y_0 X(T) \int_0^T X^{-1}(s)f(s)ds = 0$$

It follows that the rank of the matrix $I - X(T)$ is the same as the rank of the augmented matrix

$$[I - X(T) \vdots X(T) \int_0^T X^{-1}(s)f(s)ds]$$

(*) If D is a row vector, then $[C \vdots D]$ denotes a row-augmented matrix.

Hence, the system (6.4) has at least one solution x_0, and this completes the proof.

The following important result says that if the system (S_f) has at least one bounded solution on R_+, then (S_f) has at least one T-periodic solution.

Theorem 6.4 If the system (S_f) does not have any T-periodic solutions, then (S_f) has no bounded solutions.

Proof. Suppose (S_f) does not have any T-periodic solutions. Then (S_a) must have at least one nontrivial T-periodic solution. If this is not true, then Theorem 6.1 implies the existence of a T-periodic solution of (S_f), since the matrix $\tilde{X} = X(0) - X(T)$ is nonsingular. Thus, there must exist a nontrivial T-periodic solution $y(t)$ of (S_a) which satisfies

$$\int_0^T y(t)f(t)dt \neq 0$$

This follows from Theorem 6.3. Let $y(0) = y_0$. Then we have

$$y_0[I - X(T)] = 0, \quad y_0 \int_0^T X^{-1}(t)f(t)dt \neq 0 \tag{6.9}$$

because $y(t) = y_0 X^{-1}(t)$ and $y(t)$ is T-periodic. Now let $x(t)$ be any solution of (S_f) with $x(0) = x_0$. Then the variation of constants formula implies

$$x(T) = X(T)[x_0 + \int_0^T X^{-1}(s)f(s)ds] \tag{6.10}$$

which yields

$$y_0 x(T) = y_0 X(T)x_0 + y_0 X(T) \int_0^T X^{-1}(s)f(s)ds \tag{6.11}$$

$$= y_0 x_0 + y_0 \int_0^T X^{-1}(s)f(s)ds$$

We also have

$$x(t + T) = X(t)x(T) + \int_0^t X(t)X^{-1}(s)f(s)ds \tag{6.12}$$

because both members of (6.12) are solutions of (S_f) with the value $x(T)$ at $t = 0$. Thus,

$$x(2T) = X(T)[x(T) + \int_0^T X^{-1}(s)f(s)ds]$$

which, along with (6.10) and (6.11), implies

$$y_0 x(2T) = y_0 x(T) + y_0 \int_0^T X^{-1}(s)f(s)ds \tag{6.13}$$

$$= y_0 x_0 + 2y_0 \int_0^T X^{-1}(s)f(s)ds$$

Similarly, by induction, we obtain

$$y_0 x(nT) = y_0 x_0 + n y_0 \int_0^T X^{-1}(s) f(s) ds, \quad n = 1, 2, \ldots$$

It follows that $x(t)$ cannot be a bounded solution of (S_f) on R_+. In fact, if $x(t)$ was bounded on R_+, then $y_0 x(nT)$ would be bounded. This in turn would imply that the sequence

$$n y_0 \int_0^T X^{-1}(s) f(s) ds, \quad n = 1, 2, \ldots$$

is bounded, which is a contradiction to the second relation of (6.9). This completes the proof.

3. Dependence of $x(t)$ on A, U

In this section we study the dependence of the solution $x(t)$, of the problem $((S_f), B_3)$, on the matrix A and the operator U. We actually show that $x(t)$ is a continuous function of A, U in a certain sense. In what follows, U will be assumed to be a bounded linear operator on $C_n[0, T]$ with values in R^n. We should recall here that $C_n[0, T]$ is associated with the sup-norm. The norm of R^n in this section will be $\|x\| = \sum_{i=1}^{n} |x_i|$. The corresponding matrix norm is $\|A\| = \max_k \sum_i |a_{ik}|$. For a matrix $A(t)$ or a vector $x(t)$, $\|A\|$, $\|x\|$ denote the sup-norms. The proof of the following lemma is left as an exercise.

Lemma 6.5 Let $A \in M_n$ be nonsingular with $\|A^{-1}\| = M$. Then any matrix $B \in M_n$ with $\|A - B\| < m$ is nonsingular if $mM < 1$. Here m is a positive constant.

The fundamental matrix $X(t)$ of (S) with $X(0) = I$ will be denoted by $X_A(t)$. We also denote the matrix $\tilde{X}$, introduced in Section 1, by $\tilde{X}_A$. By (S_B) we denote the system (S) with $A(t)$ replaced by $B(t)$. We have Lemma 6.6.

Lemma 6.6 Let $A: [0, T] \to M_n$ be continuous. Then for every $\epsilon > 0$ there exists $\delta(\epsilon) > 0$ such that for every continuous $B: [0, T] \to M_n$ with

$$\int_0^T \|A(t) - B(t)\| dt < \delta(\epsilon)$$

we have

$$\|X_A - X_B\| < \epsilon$$

Proof. From

$$X_B'(t) = A(t) X_B(t) + [B(t) - A(t)] X_B(t)$$

it is easy to see that

$$X_B(t) = X_A(t)[I + \int_0^t X_A^{-1}(s)[B(s) - A(s)]X_B(s)ds] \tag{6.15}$$

Letting

$$K = \max_{t \in [0,T]} \|X_A(t)\|, \quad L = \max_{t \in [0,T]} \|X_A^{-1}(t)\|$$

we obtain

$$\begin{aligned}\|X_A(t) - X_B(t)\| &\leq \|X_A(t) \int_0^t X_A^{-1}(s)[A(s) - B(s)][X_A(s) - X_B(s)]ds\| \\ &+ \|X_A(t) \int_0^t X_A^{-1}(s)[A(s) - B(s)]X_A(s)ds\| \\ &\leq K^2 L \int_0^T \|A(t) - B(t)\|dt \\ &+ KL \int_0^t \|A(s) - B(s)\| \|X_A(s) - X_B(s)\|ds \end{aligned} \tag{6.16}$$

for every $t \in [0, T]$. To get this, we have added and subtracted an obvious term to the right hand side of (6.15). Applying Gronwall's inequality to (6.16), we obtain

$$\|X_A(t) - X_B(t)\| \leq C \exp\{KL \int_0^T \|A(t) - B(t)\|dt\}$$

where

$$C = K^2 L \int_0^T \|A(t) - B(t)\|dt$$

Thus, it suffices to pick as $\delta(\epsilon)$ any positive number λ such that

$$K^2 L \lambda e^{KL\lambda} < \epsilon$$

Lemma 6.7 Let A be as in Lemma 6.6 with $\tilde{X}_A$ nonsingular. Then there exists a number $\delta > 0$ such that for every continuous $B: [0, T] \to M_n$ with

$$\int_0^T \|A(t) - B(t)\|dt < \delta \tag{6.17}$$

the matrix $\tilde{X}_B$ is also nonsingular.

Proof. Let $V = \tilde{X}_A - \tilde{X}_B = [v_{ij}]$ and $X_A(t) = [x_{ij}(t)]$, $X_B(t) = [y_{ij}(t)]$. Then we have

$$\sum_{i=1}^n |v_{ij}| = \|U([x_{1j}(\cdot) - y_{1j}(\cdot), \ldots, x_{nj}(\cdot) - y_{nj}(\cdot)]^T)\|$$

$$\leq \|U\| \|[x_{1j}(\cdot) - y_{1j}(\cdot), \ldots, x_{nj}(\cdot) - y_{nj}(\cdot)]^T\|$$

$$= \|U\| \sup_{t \in [0,T]} \sum_{i=1}^{n} |x_{ij}(t) - y_{ij}(t)|$$

Consequently,

$$\|V\| = \max_{j} \sum_{i=1}^{n} |v_{ij}|$$

$$\leq \|U\| \max_{j} \sup_{t \in [0,T]} \sum_{i=1}^{n} |x_{ij}(t) - y_{ij}(t)|$$

$$= \|U\| \sup_{t \in [0,T]} \max_{j} \sum_{i=1}^{n} |x_{ij}(t) - y_{ij}(t)|$$

$$= \|U\| \sup_{t \in [0,T]} \|X_A(t) - X_B(t)\|$$

$$= \|U\| \|X_A - X_B\|$$

If we let ϵ_0 be such that $\epsilon_0 \|U\| \|\tilde{X}_A^{-1}\| < 1$, then, by Lemma 6.6, there exists $\delta(\epsilon_0) > 0$ such that

$$\int_0^T \|A(t) - B(t)\| dt < \delta(\epsilon_0) \tag{6.18}$$

implies $\|X_A - X_B\| < \epsilon_0$. This says that whenever (6.18) holds we have

$$\|V\| = \|\tilde{X}_A - \tilde{X}_B\| \leq \epsilon_0 \|U\|$$

which, by Lemma 6.5, implies the invertibility of the matrix $\tilde{X}_B$.

Now we are ready for the following important corollaries.

Corollary 6.8 Consider the system (S) with $A: [0, T] \to M_n$ continuous. Assume further that $\tilde{X}_A^{-1}$ exists. Then there exists $\delta_1 > 0$ such that for every continuous $B: [0, T] \to M_n$ with

$$\int_0^T \|A(t) - B(t)\| dt < \delta_1$$

the problem $((S_B), (B_3))$ has a unique solution for every $f \in C_n[0, T]$ and every $r \in R^n$.

Corollary 6.9 Let A, $\tilde{X}_A$ be as in Corollary 6.8. Then there exists $\delta_2 > 0$ such that for every bounded linear operator $U_1: C_n[0, T] \to R^n$ with $\|U - U_1\| < \delta_2$, the problem consisting of (S_f) and

$$U_1 x = r \tag{B_5}$$

has a unique solution for every $f \in C_n[0, T]$ and every $r \in R^n$.

Proof. Let U_1 be a bounded linear operator mapping $C_n[0, T]$ into R^n and assume that $\widetilde{X}_{1,A}$ denotes the matrix whose columns are the values of U_1 on the corresponding columns of $X_A(t)$. Moreover, let $W = \widetilde{X}_A - \widetilde{X}_{1,A} = [w_{ij}]$. Then if $X_A(t) = [x_{ij}(t)]$, we have

$$\sum_{i=1}^{n} |w_{ij}| = \|(U - U_1)(x_{1j}(\cdot), x_{2j}(\cdot), \ldots, x_{nj}(\cdot))^T)\|$$

$$\leq \|U - U_1\| \|[x_{1j}(\cdot), x_{2j}(\cdot), \ldots, x_{nj}(\cdot)]^T\|$$

$$= \|U - U_1\| \sup_{t \in [0,T]} \sum_{i=1}^{n} |x_{ij}(t)|$$

It follows that

$$\|W\| = \max_{j} \sum_{i=1}^{n} |w_{ij}|$$

$$\leq \|U - U_1\| \sup_{t \in [0,T]} \max_{j} \sum_{i=1}^{n} |x_{ij}(t)|$$

$$= \|U - U_1\| \sup_{t \in [0,T]} \|X(t)\|$$

$$= \lambda \|U - U_1\|$$

where λ is the sup-norm of $X(t)$. Thus, according to Lemma 6.5, it suffices to take $\delta_2 = \mu/\lambda$, where μ is a positive number with $\mu \|\widetilde{X}_A^{-1}\| < 1$. For such δ_2, $\widetilde{X}_{1,A}$ is nonsingular. This, by Theorem 6.1, proves our assertion.

Combining the above corollaries, we obtain the main result of this section (Theorem 6.10).

Theorem 6.10 Let $A: [0, T] \to M_n$ be continuous and let $U: C_n[0, T] \to R^n$ be a bounded linear operator such that $\widetilde{X}_A$ is nonsingular. Then there exist two positive numbers δ_1, δ_2 such that for every continuous $B: [0, T] \to M_n$ with

$$\int_0^T \|A(t) - B(t)\| dt < \delta_1$$

and every bounded linear operator $U_1: C_n[0, T] \to R^n$ with $\|U - U_1\| < \delta_2$ the problem $((S_B), (B_5))$ has a unique solution for every $f \in C_n[0, T]$ and every $r \in R^n$.

Naturally, the integral condition in Theorem 6.10 can be replaced by $\|A - B\| < \delta_1/T = \delta_1^*$.

4. Perturbed Linear Systems

In this section we are interested in the solutions of the problem

$$x' = A(t)x + F(t, x), \qquad (S_F)$$

$$Ux = r \qquad (B_3)$$

where A, U, r are as in the previous section and $F: [0, T] \times R^n \to R^n$ is continuous. We shall study the same problem on the interval R_+. As we saw in Section 1, the problem $((S_F), (B_3))$ will have a solution on $[0, T]$ if a function $x(t)$, $t \in [0, T]$, can be found satisfying the integral equation

$$x(t) = X(t)\widetilde{X}^{-1}[r - Up(\cdot, x)] + p(t, x) \qquad (6.19)$$

where

$$p(t, x) = \int_0^t X(t)X^{-1}(s)F(s, x(s))ds$$

for $t \in [0, T]$. In the following result we apply the Schauder-Tychonov theorem in order to obtain a fixed point of the operator defined, on $C_n[0, T]$, be the right hand side of (6.19). Given a number $\alpha > 0$, We shall denote by S_α the set $\{u \in R^n; \|u\| \leq \alpha\}$.

Theorem 6.11 Let

$$q(t) = \max_{u \in S_\alpha} \{\|X^{-1}(t)F(t, u)\|\}, \quad t \in [0, T]$$

and define the operator K as follows:

$$Kf = \widetilde{X}^{-1}[r - Up(\cdot, f)]$$

for every $f \in S^\alpha = \{g \in C_n[0, T]; g(t) \in S_\alpha, t \in [0, T]\}$. Suppose that $L(M + N) \leq \alpha$, where

$$L = \max_{t \in [0,T]} \{\|X(t)\|\}, \ M = \sup_{g \in S^\alpha} \{\|Kg\|\}, \ N = \int_0^T q(t)dt$$

Then the problem $((S_F), (B_3))$ has at least one solution.

Proof. We are going to show that the operator $V: S^\alpha \to C_n[0, T]$ defined by

$$(Vu)(t) = X(t)[Ku + \int_0^t X^{-1}(s)F(s, u(s))ds]$$

has a fixed point in S^α. To this end, let $t, t_1 \in [0, T]$ be given. Then we have

$$\|(Vu)(t) - (Vu)(t_1)\| \le \|X(t)[Ku + \int_0^t X^{-1}(s)F(s, u(s))ds]$$
$$- X(t_1)[Ku + \int_0^{t_1} X^{-1}(s)F(s, u(s))ds]\|$$
$$\le M\|X(t) - X(t_1)\| + N\|X(t) - X(t_1)\|$$
$$+ \|X(t_1)\| \left| \int_t^{t_1} q(s)ds \right|$$
$$\le (M + N)\|X(t) - X(t_1)\| + L\left|\int_t^{t_1} q(s)ds\right| \quad (6.20)$$

To obtain this, we have expressed the integral $\int_0^t$ in the first inequality as a sum $\int_0^{t_1} + \int_{t_1}^t$. Now let $\epsilon > 0$ be given. Then there exists $\delta(\epsilon) > 0$ such that

$$\|X(t) - X(t_1)\| < \epsilon/2(M + N), \quad \left|\int_t^{t_1} q(s)ds\right| < \epsilon/2L \quad (6.21)$$

for every $t, t_1 \in [0, T]$ with $|t_1 - t| < \delta(\epsilon)$. This follows from the uniform continuity of the function $X(t)$ and the function

$$h(t) = \int_0^t q(s)ds$$

on the interval $[0, T]$. Inequalities (6.20) and (6.21) imply the equicontinuity of the set VS^α. The fact that $VS^\alpha \subset S^\alpha$ follows from $L(M + N) \le \alpha$. Thus, VS^α is relatively compact (see Theorem 2.5). Now we show that V is continuous on S^α. In fact, let $\{u_m\}_{m=1}^\infty \subset S^\alpha$, $u \in S^\alpha$ satisfy

$$\|u_m - u\|_\infty \to 0 \text{ as } m \to \infty$$

Then we have

$$\|Vu_m - Vu\|_\infty \le L[\|Ku_m - Ku\| + \int_0^T \|X^{-1}(s)[F(s, u_m(s)) - F(s, u(s))]\|ds]$$

$$\le L(L\|\tilde{X}^{-1}\|\|U\| + 1) \int_0^T \|X^{-1}(s)[F(s, u_m(s)) - F(s, u(s))]\|ds$$
$$(6.22)$$

The integrand in the last member of (6.22) converges uniformly to zero. Thus, $\|Vu_m - Vu\|_\infty \to 0$ as $m \to \infty$. The relative compactness of S^α and the continuity of V on S^α imply, by the Schauder-Tychonov theorem, the existence of a fixed point $x(t), t \in [0, T]$, of the operator V. This function $x(t)$ is a solution to the problem $((S_f), (B_3))$.

Theorem 6.12 Let S_α, S^α be as in Theorem 6.11 for a fixed $\alpha > 0$. Assume that there exists a constant $k > 0$ such that for every $x_0 \in S_\alpha$ the solution $x(t, 0, x_0)$ of (S_F), with $x(0) = x_0$, exists on $[0, T]$, it is unique and satisfies

$$\sup_{\substack{t\in[0,T]\\x_0\in S_\alpha}} \|x(t, 0, x_0)\| \le k$$

Let

$$N = \int_0^T q(s)ds, \text{ where } q(t) = \max_{\|u\|\le k} \|X^{-1}(t)F(t, u)\|$$

Then the condition $\|\widetilde{X}^{-1}\|(\|r\| + \|U\|LN) \le \alpha$ (with L defined as in Theorem 6.11) implies the existence of a solution to the problem $((S_F), (B_3))$.

Proof. We are going to apply Brouwer's fixed point theorem (Corollary 2.15). To this end, consider the operator $Q: S_\alpha \to R^n$ defined by

$$Qu = \widetilde{X}^{-1}(r - Up_1(\cdot, u)) \tag{6.23}$$

where

$$p_1(t, u) = \int_0^t X(t)X^{-1}(s)F(s, x(s, 0, u))ds$$

It is easy to see that our asssumptions imply that $QS_\alpha \subset S_\alpha$. To show the continuity of Q, we need to show the continuity of $x(t, 0, u)$ w.r.t. u. Let $u_m \in S_\alpha$, $m = 1, 2, \ldots$, $u \in S_\alpha$ be such that $\|u_m - u\| \to 0$ as $m \to \infty$ and let $x_m(t)$, $x(t)$ be the solutions of

$$x' = A(t)x + F(t, x), \quad x(0) = u_m,$$

$$x' = A(t)x + F(t, x), \quad x(0) = u$$

respectively. Then our assumptions imply that $\|x_m(t)\| \le k$, $\|x(t)\| \le k$ for $t \in [0, T]$, $m = 1, 2, \ldots$. Hence, the functions $x_m(t)$, $m = 1, 2, \ldots$, form a uniformly bounded family. From

$$\|x_m'(t)\| \le k\sup_{t\in[0,T]} \|A(t)\| + \sup_{\substack{t\in[0,T]\\ \|u\|\le k}} \|F(t, u)\|$$

it follows that this family is also equicontinuous. By Theorem 2.5, there exists a subsequence $\{x_j(t)\}_{j=1}^\infty$ of $\{x_m(t)\}$ such that $x_j(t) \to \bar{x}(t)$ as $j \to \infty$ uniformly on $[0, T]$. Here $\bar{x}(t)$ is some function in $C_n[0, T]$. Taking limits as $j \to \infty$ in

$$x_j(t) = x_j(0) + \int_0^t A(s)x_j(s)ds + \int_0^t F(s, x_j(s))ds$$

we obtain that

$$\bar{x}(t) = u + \int_0^t A(s)\bar{x}(s)ds + \int_0^t F(s, \bar{x}(s))ds$$

Thus, by uniqueness, $\bar{x}(t) = x(t)$, $t \in [0, T]$. Since we could have started with any subsequence of $\{x_m(t)\}$ instead of $\{x_m(t)\}$ itself, we have actually shown the following: every subsequence of $\{x_m(t)\}$ contains a subsequence converging uniformly to $x(t)$ on $[0, T]$. This implies the uniform convergence of $\{x_m(t)\}$ to $x(t)$ on $[0, T]$. Equivalently, if $u_m \in S_\alpha$, $m = 1, 2, \ldots, u \in S_\alpha$ satisfy $\|u_m - u\| \to 0$ as $m \to \infty$, then $\|x(t, 0, u_m) - x(t, 0, u)\| \to 0$ uniformly on $[0, T]$. This yields the continuity of the function $x(t, 0, x_0)$ w.r.t. $x_0 \in S_\alpha$ uniformly in $t \in [0, T]$. Now let $\{u_m\}$, u be as above. We have

$$\|Qu_m - Qu\| \leq \|X^{-1}\| \|U\| L \int_0^T \|X^{-1}(s)[F(s, x(s, 0, u_m))$$

$$- F(s, x(s, 0, u))]\| ds \qquad (6.24)$$

The integrand in (6.24) tends uniformly to 0 as $m \to \infty$. Thus $\|Qu_m - Qu\| \to 0$ as $m \to \infty$, and this implies the continuity of Q on S_α. Brouwer's fixed point theorem implies now the existence of some $x_0 \in S_\alpha$ with the property $Qx_0 = x_0$. This vector x_0 is the initial value of a solution to the problem $((S_F), (B_3))$.

5. Problems on Infinite Intervals

Theorem 6.11 can be extended to problems on infinite intervals. Actually, in this section we show that this can be done in two ways. If we work in the space $C_n(R_+)$, we "truncate" the problem, by solving an equation such as (6.19) on $[0, m]$, $m = 1, 2, \ldots$, and then obtain a solution on R_+ by approximation. If the domain of the operator V is C_n^ℓ, then we can solve (6.19) by finding directly a fixed point of V. This second method of solution is possible because we can detect the compact sets in C_n^ℓ (see Exercise 2.5).

The following lemma is useful in the sequel. Its proof is left as an exercise.

Lemma 6.13 (Lebesgue's dominated convegence theorem). Let $f_m : [t_0, \infty) \to R^n$, $m = 1, 2, \ldots$, $t_0 \geq 0$, be continuous and such that the improper integrals

$$\int_{t_0}^\infty f_m(t) dt, \quad m = 1, 2, \ldots$$

are convergent. Let $f_m \to f$ as $m \to \infty$ pointwise on $[t_0, \infty)$, where $f : [t_0, \infty) \to R^n$ is continuous and such that the improper integral

$$\int_{t_0}^\infty f(t) dt$$

is convergent. Furthermore, let $\|f_m(t)\| \leq g(t)$, $t \in [t_0, \infty)$, $m = 1, 2, \ldots$, where $g : [t_0, \infty) \to R_+$ is continuous and such that

$$\int_{t_0}^\infty g(t) dt < +\infty$$

Then we have

$$\lim_{m\to\infty} \int_{t_0}^{\infty} f_m(t)dt = \int_{t_0}^{\infty} \lim_{m\to\infty} f_m(t)dt = \int_{t_0}^{\infty} f(t)dt$$

In the proof of the next theorem the integral operator V is defined on $C_n(R_+)$. The symbol $\tilde{X}$ will denote again the matrix in M_n whose columns are the values of U on the corresponding columns of $X(t)$. This of course means that the columns of $X(t)$ belongs to $C_n(R^+)$. A similar situation will be assumed for the space C_n^l.

Theorem 6.14 Let $A: R_+ \to M_n$, $F: R_+ \times R^n \to R^n$ be continuous and such that $\|X(t)\| \le M$, $t \in R_+$, where M is a positive constant. Furthermore, assume the following:

(i) there exist two continuous function $q, g: R_+ \to R_+$ such that

$$\|X^{-1}(t)F(t, u)\| \le q(t)\|u\| + g(t), \quad (t, u) \in R_+ \times R^n \quad (6.25)$$

and

$$L = \int_0^{\infty} q(t)dt < +\infty, \quad N = \int_0^{\infty} g(t)dt < +\infty;$$

(ii) $U: C_n(R_+) \to R^n$ is a bounded linear operator such that $\tilde{X}^{-1}$ exists.

(iii) $LM^2 \|\tilde{X}^{-1}\| \|U\| e^{ML} < 1$.

Then the problem $((S_F), (B_3))$ has at least one solution.

Proof. As in Theorem 6.11, it suffices to show that the equation

$$x(t) = X(t)\tilde{X}^{-1}[r - Up(\cdot, x)] + p(t, x) \quad (6.19)$$

has a solution $x \in C_n(R_+)$. We consider the spaces D_m, $m = 1, 2, \ldots$, defined as follows:

$$D_m = \{x \in C_n(R_+); x(t) = x(m), t \ge m\}$$

The spaces D_m are Banach spaces with norms

$$\|x\|_m = \sup_{t \in [0,m]} \|x(t)\|$$

We define the operator $V_1: D_1 \times [0, 1] \to D_1$ as follows: if $(f, \mu) \in D_1 \times [0, 1]$, then $V_1(f, \mu) = h \in D_1$, where the function h has a restriction $\bar{h}$ on $[0, 1]$ satisfying

$$\bar{h}(t) = \mu X(t)\tilde{X}^{-1}[r - Up(\cdot, f)] + \mu p(t, f), \quad t \in [0, 1]$$

We are planning to apply the Leray-Schauder theorem in order to obtain a fixed point for the operator $V_1(x, 1)$ in D_1. We first show that $V_1(x, \mu)$ is continuous in x. Let $\{f_k\}_{k=1}^{\infty} \subset D_1$, $f \in D_1$, be given with $f_k \to f$ as $k \to \infty$ uniformly on R_+. This is equivalent to saying that

$$\|f_k - f\|_1 \to 0 \text{ as } k \to \infty$$

Then we have

$$\|V_1(f_k, \mu) - V_1(f, \mu)\|_1$$
$$\leq M^2 \|\widetilde{X}^{-1}\| \|U\| \int_0^{\infty} \|X^{-1}(s)[F(s, f_k(s)) - F(s, f(s))]\| ds$$
$$+ M \int_0^{\infty} \|X^{-1}(s)[F(s, f_k(s)) - F(s, f(s))]\| ds \qquad (6.26)$$

Since $\|F(t, f_k(t)) - F(t, f(t))\| \to 0$ as $m \to \infty$ pointwise on R_+ and

$$\|X^{-1}(t)[F(t, f_k(t)) - F(t, f(t))]\|$$
$$\leq q(t)[\|f_k\|_1 + \|f\|_1 + \|f\|_1] + 2g(t), \quad t \in R_+$$

Inequality 6.26, Lemma 6.13 and our hypotheses on q, g imply that

$$\|V_1(f_k, \mu) - V_1(f, \mu)\|_1 \to 0 \text{ as } k \to \infty$$

It follows that $V_1(\cdot, \mu)$ is continuous on D_1. The proof of the equicontinuity of the set $\{V_1(f, \mu); f \in B\}$, for any bounded set $B \subset D_1$ and fixed μ, follows easily, as in Theorem 6.11, and is therefore omitted. In order to show that all possible solutions of $V_1(x, \mu) = x$ are in a ball of D_1 that does not depend on μ, let $x \in D_1$ solve $V_1(x, \mu_0) = x$ for some $\mu_0 \in [0, 1]$. Then

$$\|x(t)\| \leq M\|\widetilde{X}^{-1}\|\{\|r\| + M\|U\|(L\|x\|_1 + N)\}$$
$$+ M \int_0^t q(s)\|x(s)\| ds + MN, \quad t \in [0, 1] \qquad (6.27)$$

Letting

$$K = LM^2\|\widetilde{X}^{-1}\|\|U\|, \quad Q = M\|\widetilde{X}^{-1}\|(\|r\| + M\|U\|N) + MN$$

we have

$$\|x(t)\| \leq K\|x\|_1 + Q + M \int_0^t q(s)\|x(s)\| ds, \quad t \in [0, 1]$$

Applying Gronwall's inequality, we arrive at

$$\|x(t)\| \leq (K\|x\|_1 + Q) \exp\{M \int_0^t q(s)ds\}$$

$$\leq (K\|x\|_1 + Q)e^{ML}, \quad t \in [0, 1] \tag{6.28}$$

which yields

$$\|x\|_1 \leq (1 - Ke^{ML})^{-1} Q e^{ML} \tag{6.29}$$

Consequently, every solution $x \in D_1$ of $V_1(x, \mu) = x$ satisfies $\|x\|_1 \leq \alpha$, where the constant α equals the right hand side of (6.29) and is independent of μ. The fact that $V_1(x, \mu)$ maps bounded sets into bounded sets for each $\mu \in [0, 1]$ follows from

$$\|V_1(f, \mu)\|_1 \leq K\|f\|_1 + Q + M \int_0^\infty q(t)dt \|f\|_1$$

which can be obtained as the inequality preceding (6.28). This property and the equicontinuity of the image under V_1 of any bounded subset of D_1, for fixed μ, imply the compactness of $V_1(\cdot, \mu)$ for each $\mu \in [0, 1]$. The uniform continuity of $V_1(x, \mu)$ in μ w.r.t. x in bounded subsets of D_1 is an easy consequence of the definition of V_1. Using the Leray-Schauder theorem, we obtain a solution $x_1 \in S_\alpha = \{x \in C_n[0, \infty); \|\cdot\|_\infty \leq \alpha\}$ of the equation $V_1(x, 1) = x$. This solution belongs to D_1 and satisfies (6.19) on $[0, 1]$. Continuing the same process, we obtain a sequence of solutions $x_m \in D_m$, $m = 1, 2, \ldots$, such that each $x_m(t)$ satisfies (6.19) on $[0, m]$ and belongs to S_α.

It is easy to see now that the sequence $\{x_m(t)\}_{m=1}^\infty$ is equicontinuous on the interval $[0, 1]$. By Theorem 2.5, there exists a subsequence $\{x_m^1(t)\}_{m=1}^\infty$ of $\{x_m(t)\}$ such that $x_m^1(t) \to x^1(t)$ as $m \to \infty$ uniformly on $[0, 1]$. Similarly, there exists a subsequence $\{x_m^2(t)\}_{m=1}^\infty$ of $\{x_m^1(t)\}$ converging uniformly to a function $x^2(t)$, $t \in [0, 2]$, such that $x^2(t) = x^1(t)$, $t \in [0, 1]$ and so on. The diagonal sequence $\{x_m^m(t)\}_{m=1}^\infty$ is a subsequence of the original sequence such that $x_m^m(t) \to x(t)$ uniformly on every interval $[0, c]$, $c > 0$. Here $x(t)$ is a continuous function on R_+ with $\|x(t)\| \leq \alpha$ for $t \in R_+$. Let

$$y(t) = X(t)X^{-1}[r - Up(\cdot, x)] + p(t, x), \quad t \in R_+$$

Then, for any $c > 0$ and $m \geq c$, we get

$$\|x_m^m(t) - y(t)\|$$
$$\leq M(M\|\tilde{X}^{-1}\|\|U\| + 1) \int_0^\infty \|X^{-1}(s)[F(s, x_m^m(s)) - F(s, x(s))]\| ds, \quad t \in [0, c]$$

Applying once more the Lebesgue dominated convergence theorem (Lemma 6.13), we obtain that $x_m^m \to y$ as $m \to \infty$ uniformly on every interval $[0, c]$. This shows that $x(t) \equiv y(t)$, $t \in R_+$. The function $x(t)$ is a solution to the problem $((S_F), (B_3))$.

In Theorem 6.15 the space $C_n(R_+)$ is replaced by C_n^l.

Theorem 6.15 Along with the assumptions (i) - (iii) of Theorem 6.14, assume that

$$\lim_{t \to \infty} X(t) = X(\infty)$$

exists as a finite matrix. Furthermore, assume that $U: C_n^l \to R^n$ is a bounded linear operator such that $\widetilde{X}^{-1}$ exists.

Then, for every $r \in R^n$, the problem $((S_F), (B_3))$ has at least one solution.

Proof. We consider the operator $V(x, \mu): C_n^l \times [0, 1] \to C_n^l$ defined as follows:

$$(V(f, \mu))(t) = \mu X(t)\widetilde{X}^{-1}[r - Up(\cdot, f)] + \mu p(t, f), \quad t \in R_+$$

In order to apply the Leray-Schauder theorem, we show here only the equiconvergence (Exercise 2.5, (iii)) of the set

$$\{V(f, \mu); f \in S_\alpha\} \tag{6.30}$$

for every $\mu \in [0, 1]$, where S_α is given in the proof of Theorem 6.14. The rest of the assumptions of the Leray-Schauder theorem follow as in Theorem 6.14 and are therefore omitted.

Let $f \in S_\alpha$ be given, and let

$$\lim_{t \to \infty} (V(f, \mu))(t) = \xi$$

for some $\mu \in [0, 1]$. Then we have

$$\|V(f, \mu)(t) - \xi\| \leq \|X(t) - X(\infty)\| \|\widetilde{X}^{-1}\| \|r\|$$

$$+ M\|X(t) - X(\infty)\| \|\widetilde{X}^{-1}\| \|U\|(L\|f\|_\infty + N)$$

$$+ \|X(t) - X(\infty)\| \int_0^\infty \|X^{-1}(s)F(s, f(s))\| ds$$

$$+ \|X(\infty)\| \int_t^\infty \|X^{-1}(s)F(s, f(s))\| ds$$

$$\leq \|X(t) - X(\infty)\| \|\widetilde{X}^{-1}\| \|r\|$$

$$+ M\|X(t) - X(\infty)\| \|\widetilde{X}^{-1}\| \|U\|(L\alpha + N)$$

$$+ \|X(t) - X(\infty)\|(L\alpha + N)$$

$$+ \alpha M \int_t^\infty q(s)ds + M \int_t^\infty g(s)ds$$

This proves the equiconvergence of the set in (6.30). By the Leray-Schauder theorem, $V(x, 1)$ has at least one fixed point in S_α.

Remark 6.16 Naturally, the results of the last two sections, concerning the existence of solutions of boundary value problems of perturbed linear systems, have corresponding versions for contraction integral operators. This is the content of Exercise 6.7, where extensions are sought for Theorems 6.14 and 6.15.

Example 6.17 Consider the system

$$\begin{bmatrix} x_1 \\ x_2 \end{bmatrix}' = \begin{bmatrix} 0 & 1 \\ -1 & 0 \end{bmatrix} \begin{bmatrix} x_1 \\ x_2 \end{bmatrix} + \begin{bmatrix} 0 \\ \epsilon e^{-t} \ln(|x_1| + 1) + [1/(t^2 + 1)] \cdot \sin x_1 \end{bmatrix}$$

(6.31)

and the boundary conditions

$$Ux = \int_0^\infty P(t)x(t)dt = \begin{bmatrix} -1 \\ \pi \end{bmatrix}$$

(6.32)

for $x \in C_n(R_+)$, where

$$P(t) = \begin{bmatrix} e^{-t} & -e^{-3t} \\ 0 & 2e^{-2t} \end{bmatrix}$$

Choosing, for convenience, the norm $\|x\| = |x_1| + |x_2|$ in R^2, we have $\|A\| = \sup_k \sum_i |a_{ik}|$ for the 2×2 matrix A. Thus, we obtain

$$\int_0^\infty \|P(t)dt\| < +\infty$$

and

$$X(t) = \begin{bmatrix} \cos t & \sin t \\ -\sin t & \cos t \end{bmatrix}, \quad \tilde{X} = \begin{bmatrix} 3/5 & 1/5 \\ -2/5 & 4/5 \end{bmatrix}$$

We also find

$$\|X^{-1}(t)F(t, u)\| \leq 2\epsilon e^{-t}\|u\| + 2/(t^2 + 1)$$

It follows from Theorem 6.14 that the problem ((6.31), (6.32)) has at least one solution for all sufficiently small $\epsilon > 0$.

Theorem 7.14 contains another application of the Leray-Schauder theorem to b.v.p.'s on infinite intervals.

EXERCISES

6.1. For the system (S_f) assume that $A: R \to M_n$, $f: R \to R^n$ are continuous and T-periodic. Assume further that the only T-periodic solution of the homogeneous system (S) is the zero solution. Find a function $G(t, s)$ such that

$$x(t) = \int_0^T G(t, s)f(s)ds$$

where $x(t)$ is the (unique) T-periodic solution of (S_f).

6.2. Assume that $A: R \to M_n$ is continuous and T-periodic. Assume further that the system (S_f) has at least one T-periodic solution for any $f \in C_n(R)$ which is T-periodic. Show that the homogeneous system (S) has a unique T-periodic solution — the zero solution.

6.3. Consider the scalar equation

$$x'' - x = f(t) \qquad (6.33)$$

where $f: [0, T] \to R$ is continuous. Show that this equation has a unique solution $x(t)$, $t \in [0, T]$, such that

$$x(0) = x(T), \quad x'(0) = x'(T) \qquad (6.34)$$

[Hint. Examine the system in R^2 arising from (6.33).]

6.4. Let $F: [0, T] \times R \to R$ be continuous and such that $|F(t, u)| \leq M$ for $t \in [0, T]$, $|u| \leq \alpha$, where α is a positive constant. Show that if M is sufficiently small, then the equation

$$x'' - x = F(t, x)$$

has at least one solution $x(t)$, $t \in [0, T]$, satisfying (6.34). [Hint. Apply the Schauder-Tychonov theorem.]

6.5. Consider the system

$$\begin{bmatrix} x_1 \\ x_2 \end{bmatrix}' = \begin{bmatrix} 0 & 1 \\ 1 & 0 \end{bmatrix} \begin{bmatrix} x_1 \\ x_2 \end{bmatrix} \qquad (6.35)$$

and the conditions

$$\begin{bmatrix} x_1(0) \\ x_2(0) \end{bmatrix} = \begin{bmatrix} x_1(3) \\ x_2(3) \end{bmatrix}$$

Find a constant $\delta > 0$ such that for every $B: [0, 3] \to M_n$, continuous and such that

$$\|B - A\| < \delta$$

($\|\cdot\|$ denotes sup-norm) the problem

$$x' = B(t)x, \quad x(0) = x(3) \tag{6.36}$$

has a unique solution. Here

$$A = \begin{bmatrix} 0 & 1 \\ 1 & 0 \end{bmatrix}$$

6.6. In the scalar equation

$$x'' = p(t)g(x) \tag{6.37}$$

let $p: R \to R_+ \setminus \{0\}$, $g: R \to R$ be continuous. Assume further that p is T-periodic and $ug(u) > 0$ for $u \neq 0$. Show that the only T-periodic solution of (6.37) is the zero solution. [Hint. Show that (6.37) has no nontrivial solutions with arbitrarily large zeros.]

6.7. Using the contraction mapping principle, obtain unique solutions to the b.v.p.'s considered in Theorems 6.14 and 6.15. Naturally, suitable Lipschitz conditions are needed here for the function $F(t, u)$.

6.8. Consider the scalar equation

$$x' = x + q(t) \tag{6.38}$$

where $q: R \to R$ is continuous and T-periodic. Show that the unique T-periodic solution $x(t)$ of (6.38) is given by the formula

$$x(t) = [e^T/(1 - e^T)] \int_0^T e^{t-s} q(s) ds + \int_0^t e^{t-s} q(s) ds, \quad t \in [0, T]$$

Let $T = 2\pi$ and determine the constant $k > 0$ so that the equation

$$x' = x + k(\sin t)[|x| + 1]$$

has at least one 2π-periodic solution. Extend this result to the equation

$$x' = x + f(t, x)$$

for a suitable function $f(t, u)$ which is T-periodic in t.

6.9. Prove Lemma 6.5.

6.10. Prove Lebesgue's dominated convergence theorem (Lemma 6.13).

6.11. The linear system (S_f), with $A: [0, T] \to M_n$, $f: [0, T] \to R^n$ continuous, has a unique solution satisfying the "boundary conditions"

$$Ux = r$$

where $Ux = x(0)$, for any $x \in C_n[0, T]$, and $r \in R^n$ is fixed. Consider the boundary conditions

$$U_1 x = x(0) - Nx(T) = r$$

where N is a matrix in M_n. Show that if $\|N\|$ is sufficiently small, the problem $((S_F),(U_1))$ has a unique solution.

6.12. Find constants a, b, c, d, not all zero, such that the problem

$$\begin{bmatrix} x_1 \\ x_2 \end{bmatrix}' = \begin{bmatrix} 0 & 1 \\ 1 & 0 \end{bmatrix} \begin{bmatrix} x_1 \\ x_2 \end{bmatrix}$$

$$\begin{bmatrix} x_1(0) \\ x_2(0) \end{bmatrix} - \begin{bmatrix} a & b \\ c & d \end{bmatrix} \begin{bmatrix} x_1(1) \\ x_2(1) \end{bmatrix} = \begin{bmatrix} 0 \\ 0 \end{bmatrix}$$

has a unique solution $x(t)$, $t \in [0, 1]$. [Hint. Use the result of Exercise 6.11.]

6.13 Assume that $A: R \to M_n$, $f: R \to R^n$ are continuous and T-periodic. Assume further that the fundamental matrix $X(t)$ ($X(0) = I$) of the system (S) satisfies

$$\|X(t)\| \leq e^{-\lambda t}, \quad t \in [0, T]$$

where λ is a positive constant. Show that the system (S_f) has a unique T-periodic solution.

6.14. Show that the continuity of the operator $V_1(x, \mu)$ (in the proof of Theorem 6.14) w.r.t. μ is uniform w.r.t. x, provided that x belongs to a bounded subset of D_1. Prove an analogous statement for the operator $V(x, \mu)$ of Theorem 6.15.

6.15. Consider the system

$$x' = A(t)x + F(t, x) \tag{S_F}$$

where

$$A = \begin{bmatrix} -1 & 0 \\ 0 & -1 \end{bmatrix}, \quad F(t, u) = \epsilon e^{-3t} \begin{bmatrix} \sin u_1 \\ u_2 \end{bmatrix} + \begin{bmatrix} 0 \\ e^{-3t} \sin u_1 \end{bmatrix}$$

and the boundary conditions

$$x(0) - x(\infty) = \begin{bmatrix} 0 \\ 1 \end{bmatrix} \tag{B}$$

Show that the problem $((S_F), (B))$ has at least one solution for all sufficiently small $\epsilon > 0$.

6.16. Let $F: R^2 \to R$ be continuous. Assume further that $F(t, u)$ is T-periodic in t and that for every compact set $K \subset R^2$ there exists a constant $L_K > 0$ such that

$$|F(t, u_1) - F(t, u_2)| \leq L_K |u_1 - u_2|$$

for every $(t, u_1), (t, u_2) \in K$. Let $x(t)$, $t \in R_+$, be a solution of

$$x' = F(t, x) \tag{E}$$

such that $\|x\|_\infty \leq M$, where M is a positive constant. Show that (E) has at least one T-periodic solution. [Hint. Assume that $x(t)$ is not T-periodic. Show first that the sequence $\{x_m\}_{m=1}^\infty$ with $x_m = x(mT)$ is either increasing or decreasing. Let $x(mT) \to \xi$ as $m \to \infty$. Show that $x_m(t) \to \bar{x}(t)$, $t \in R_+$, uniformly on every compact subset of R_+, where $x_m(t) = x(t + mT)$, $m = 1, 2, \ldots$, and $\bar{x}(t)$ is some continuous function on R_+. Then from

$$\bar{x}(T) = \lim_{m \to \infty} x_m(T) = \lim_{m \to \infty} x_{m+1}(0) = \xi = \bar{x}(0)$$

conclude that $\bar{x}(t + T) = \bar{x}(t)$, $t \in R_+$. The uniqueness of solutions w.r.t. initial conditions plays an important role in this proof.]

CHAPTER SEVEN

MONOTONICITY IN R^n AND DIFFERENTIAL SYSTEMS

This chapter is devoted to the study of systems of the form

$$x' = A(t)x + F(t, x) \tag{S_F}$$

under "monotonicity" assumptions on the matrices $A(t)$ and the functions $F(t, u)$. By monotonicity assumptions we mean conditions that include inner products involving $A(t)$ or $F(t, u)$. Although there is some overlapping between this chapter and Chapter Five, the present development is preferred because it constitutes a good step toward the corresponding theory in Banach or other spaces. This theory encompasses a major part of the modern theory of ordinary and partial differential equations.

In Section 1, we introduce a new norm of R^n which depends on a positive definite matrix. We also establish some fundamental properties of this norm.

In Section 2 we examine the stability properites of solutions of (S_F) via monotonicity conditions.

A type of partial stability, for solutions with initial conditions in specific regions, is the content of Section 3.

Section 4 is concerned with periodic solutions of differential systems, while Section 5 indicates the applicability of monotonicity assumptions to a

124 CHAPTER SEVEN

certain boundary value problem. This problem has boundary conditions that do not contain periodicity conditions as a special case.

1. A More General Inner Product For R^n

Using a positive definite matrix $V \in M_n$, we introduce another inner product for R^n which is reduced to the usual one if $V = I$. We have Lemma 7.1.

Lemma 7.1 Let $V \in M_n$ be positive definite and define $<\cdot, \cdot>_V$ by

$$<x, y>_V = <Vx, y>, \quad x, y \in R^n \qquad (7.1)$$

Then $<\cdot, \cdot>_V$ has the following properties:

(a) $<x, y>_V = <y, x>_V, \, x, y \in R^n$;

(b) $<x, \alpha y + \beta z>_V = \alpha <x, y>_V + \beta <x, z>_V, \, \alpha, \beta \in R, \, x, y, z \in R^n$;

(c) $<x, x>_V \geq 0, \, x \in R^n$, and $<x, x>_V = 0$ if and only if $x = 0$;

(d) $|<x, y>_V| \leq \|V\| \|x\| \|y\|, \, x, y \in R^n$.

The proof is left as an exercise.

The following lemma is needed for the establishment of a certain monotonicity property of a matrix whose eigenvalues have negative real parts. This property is the content of Lemma 7.3.

Lemma 7.2 Let $A, B \in M_n$ be given, with A having all of its eigenvalues with negative real parts and B positive definite. Then there exists a positive definite $V \in M_n$ such that

$$A^T V + VA = -B \qquad (7.2)$$

Proof. From Theorem 4.6 we know that the system $x' = Ax$ is asymptotically stable. Since, for autonomous systems, asymptotic stability is equivalent to uniform asymptotic stability, Inequality 4.5 implies that $\|e^{tA}\| \leq Ke^{-\alpha t}, \, t \in [0, \infty)$, where α, K are positive constants. Similarly, since e^{tA^T} satisfies the system

$$X' = XA^T(t)$$

we also obtain $\|e^{tA^T}\| \leq K_1 e^{-\alpha_1 t}$ for some positive constants α_1, K_1. Consequently, the matrix

$$V = \int_0^\infty e^{tA^T} B e^{tA} dt \qquad (7.3)$$

is well defined. Let $W = e^{tA^T} B e^{tA}$, $t \in R_+$. Then $W(0) = B$ and

$$W' = A^T W + WA \qquad (7.4)$$

Integrating (7.4) from 0 to ∞, taking into consideration that $W(t) \to 0$ as $t \to +\infty$, we obtain

$$-B = A^T V + VA$$

The positive definiteness of V follows easily from (7.3). This completes the proof.

It should be remarked here that V is the unique solution of (7.2), but this fact will not be needed in the sequel.

Lemma 7.3 Let A, B, V be as in Lemma 7.2. Then

$$< Ax, x >_V \le -(\lambda/2\mu) < x, x >_V$$

where λ is the smallest eigenvalue of B and μ is the largest eigenvalue of V.

Proof. We have

$$< Ax, x >_V = < VAx, x > = < Ax, V^T x >$$
$$= < Ax, Vx > = < x, A^T Vx >$$
$$= < A^T Vx, x > = - < Bx, x > - < VAx, x >$$
$$= - < Bx, x > - < Ax, x >_V \qquad (7.5)$$

for every $x \in R^n$, which implies

$$< \dot{A}x, x >_V = -(1/2) < Bx, x > \le -(\lambda/2) < x, x >$$
$$\le -(\lambda/2\mu) < x, x >_V$$

Here we have used Theorem 1.14.

The inner product $< \cdot, \cdot >_V$ induces the norm $\|\cdot\|_V$ on R^n with $\|u\|_V^2 = < Vu, u >$. Since all norms of R^n are equivalent, it is easy to see that a vector valued function $(x_1(t), \ldots, x_n(t))$ is differentiable w.r.t. the Euclidean norm if and only if it is differentiable w.r.t. to the norm $\|\cdot\|_V$.

Lemma 7.4 Let $x: R_+ \to R^n$ be continuously differentiable on $[0, T)$, $0 < T \leq +\infty$. Then $\|x(t)\|_V^2$ is also differentiable on $[0, T)$ and

$$(d/dt)\|x(t)\|_V^2 = 2 < x'(t), x(t) >_V, \quad t \in [0, T) \tag{7.6}$$

Here V is any positive definite matrix in M_n.

Proof. We have

$$(d/dt) < x(t), x(t) >_V = <(Vx(t))', x(t)> + <Vx(t), x'(t)>$$

$$= <Vx'(t), x(t)> + <Vx(t), x'(t)>$$

$$= 2 < x'(t), x(t) >_V \tag{7.7}$$

Now we state a theorem which establishes an upper bound for $|\lambda - \lambda'|$, where λ is an eigenvalue of a matrix $A \in M_n$ and λ' is a corresponding eigenvalue of another matrix $B \in M_n$. The proof of this theorem can be found in Ostrowski [33, p. 283].

Theorem 7.5 Let $A = [a_{ij}], B = [b_{ij}], i, j = 1, 2, \ldots, n$, be two real matrices with eigenvalues denoted by λ, λ' respectively. Let

$$M = \max_{i, j = 1, 2, \ldots, n} \{|a_{ij}|, |b_{ij}|\}, \tag{7.8}$$

$$(1/(nM)) \sum_{i, j=1}^{n} |a_{ij} - b_{ij}| = r \tag{7.9}$$

Then to every eigenvalue λ' corresponds an eigenvalue λ such that

$$|\lambda' - \lambda| \leq (n + 2)Mr^{1/n} \tag{7.10}$$

2. Stability of Differential Systems

In our first stability result we need the following existence, uniqueness and continuation theorem.

Theorem 7.6 Assume that $A: R_+ \to M_n$, $F: R_+ \times R^n \to R^n$ are continuous. Furthermore, assume the existence of a function $p: R_+ \to R$, continuous and such that

$$< F(t, x) - F(t, y), x - y > \leq p(t)\|x - y\|^2 \tag{7.11}$$

for every $t \in R_+$, $(x, y) \in R^n \times R^n$. Then for every $x_0 \in R^n$ there exists a unique solution $x(t)$, $t \in R_+$, of (S_F) such that $x(0) = x_0$. Moreover, if $x(t), y(t), t \in R_+$, are two solutions of (S_F) such that $x(0) = x_0$, $y(0) = y_0$, then

$$\|x(t) - y(t)\| \leq \|x_0 - y_0\| \exp\{\int_0^t (p(s) + q(s))ds\} \tag{7.12}$$

for $t \in R_+$, where $q(t)$ is the largest eigenvalue of $(1/2)[A(t) + A^T(t)]$.

Proof. Let $u(t) = x(t) - y(t)$, $t \in [0, T)$, where $x(t), y(t)$ are two solutions of (S_F) such that $x(0) = x_0$, $y(0) = y_0$, and T is some finite positive number. Then we have

$$u'(t) = A(t)u(t) + F(t, x(t)) - F(t, y(t))$$

for $t \in [0, T)$ and $u(0) = x_0 - y_0$. Applying Lemma 7.4 for $V = I$ we get

$$(d/dt)\|u(t)\|^2 = 2 < A(t)u(t) + F(t, x(t)) - F(t, y(t)), u(t) > \quad (7.13)$$

It is easy to see now that

$$2 < A(t)u(t), u(t) > = < (A(t) + A^T(t))u(t), u(t) >$$

The continuity of the symmetric matrix $(1/2)[A(t) + A^T(t)]$ implies the continuity of its largest eigenvalue $q(t)$ (see Exercise 1.22), and we have

$$2 < A(t)u(t), u(t) > \leq 2q(t)\|u(t)\|^2$$

This inequality, combined with (7.13), yields

$$(d/dt)\|u(t)\|^2 \leq 2q(t)\|u(t)\|^2 + 2p(t)\|u(t)\|^2 = 2(p(t) + q(t))\|u(t)\|^2 \quad (7.14)$$

for every $t \in [0, T)$. Applying Lemma 4.11 to (7.14), we obtain

$$\|u(t)\|^2 \leq \|u(0)\|^2 \exp\{2 \int_0^t (p(s) + q(s))ds\}, \quad t \in [0, T) \quad (7.15)$$

If $y(t)$ is a solution of (S_F) with $y(0) = 0$, defined on a right neighborhood of 0, then

$$(d/dt)\|y(t)\|^2 = 2 < A(t)y(t), y(t) > + 2 < F(t, y(t)) - F(t, 0), y(t) >$$
$$+ 2 < F(t, 0), y(t) >$$

$$\leq 2p(t)\|y(t)\|^2 + 2q(t)\|y(t)\|^2 + 2\|F(t, 0)\|\,\|y(t)\|$$

$$\leq 2(p(t) + q(t))\|y(t)\|^2 + 2\|F(t, 0)\|\,\|y(t)\|$$

Using

$$v' = 2|p(t) + q(t)|v + 2\|F(t, 0)\|\,|v|^{1/2}$$

as a comparison scalar equation (see Exercise 7.5), we find that $y(t)$ is extendable to $+\infty$ (see Theorem 5.16). If we let this function $y(t)$ in (7.15) we obtain

$$\|x(t)\| \leq \|y(t)\| + \|x(0)\| \exp\{\int_0^t (p(s) + q(s))ds\}$$

It follows that

$$\limsup_{t \to T^-} \|x(t)\| \leq \|y(T)\| + \|x(0)\| \exp\{\int_0^T (p(s) + q(s))ds\}$$

which, by Theorem 5.15, implies the continuability of every solution $x(t)$ to $+\infty$.

Inequality (7.15) implies of course the uniqueness of solutions of (S_F) w.r.t. initial conditions a 0. A similar inequality from t_0 to $t \geq t_0$ proves the uniqueness of solutions w.r.t. any initial conditions.

We note here that (7.12) with $y(t) \equiv 0$ ensures the continuation (but not the uniqueness) of all local solutions of (S_F) to $+\infty$, if $A(t)$ and $F(t, x)$ satisfy the following condition.

Condition (M) $A: R_+ \to M_n$, $F: R_+ \times R^n \to R^n$ are continuous, $F(t, 0) = 0$, $t \in R_+$, and

$$< F(t, x), x > \leq p(t)\|x\|^2, \quad x \in R^n, \; t \in R_+ \tag{7.16}$$

where $p: R_+ \to R$ is a continuous function.

Condition (M) is important in the following stability result.

Theorem 7.7 Let the condition (M) be satisfied and assume that $q(t)$ is as in Theorem 7.6. Then the zero solution of System (S_F) is stable if

$$\sup_{t \in R_+} \int_0^t [p(s) + q(s)]ds < +\infty$$

It is asymptotically stable if

$$\lim_{t \to \infty} \int_0^t [p(s) + q(s)]ds = -\infty$$

It is uniformly stable if

$$p(t) + q(t) \leq 0, \; t \in R_+$$

It is uniformly asymptotically stable if

$$p(t) + q(t) \leq -c, \; t \in R_+$$

where c is a positive constant.

Proof. The proof is almost identical to the proof of Theorem 4.12 and is left as an exercise.

In the following theorem we make use of the inner product $\langle \cdot, \cdot \rangle_V$.

Theorem 7.8 Let $A: R_+ \to M_n$ be continuous and such that $A(t) \to A_0$ as $t \to +\infty$. Here A_0 has all of its eigenvalues with negative real parts. Let B be a fixed positive definite matrix and V as in Lemma 7.2 with $A = A_0$. Let the rest of the assumptions of Condition (M) be satisfied with $\langle F(t, x), Vx \rangle$ replacing $\langle F(t, x), x \rangle$. Let q be the smallest eigenvalue of $(1/2)B$ and assume that, for some $T \geq 0$,

$$p = \sup_{t \geq T} p(t) < q$$

Then the zero solution of (S_F) is asymptotically stable.

Proof. From Lemma 7.4 we obtain

$$(d/dt)\|x(t)\|_V^2 = 2\langle A(t)x(t), Vx(t) \rangle + 2\langle F(t, x(t)), V(x(t)) \rangle \quad (7.17)$$

where $x(t)$ is a solution of (S_F) with initial condition $x(0) = x_0$. Letting $B(t) = -[A^T(t)V + VA(t)]$, we see that $B(t)$ is continuous and that $B(t) \to B$ as $t \to +\infty$, where B is the matrix in the statement of the theorem. If $q(t)$ is the smallest eigenvalue of $(1/2)B(t)$, then, as in (7.5), we have

$$2\langle A(t)x(t), Vx(t) \rangle = -\langle x(t), B(t)x(t) \rangle$$
$$\leq -2q(t)\|x(t)\|^2, \quad t \in R_+ \quad (7.18)$$

This inequality is now combined with (7.17) to get

$$(d/dt)\|x(t)\|_V^2 \leq -2(q(t) - p(t))\|x(t)\|^2, \quad t \in R_+ \quad (7.19)$$

Naturally, the extendability of the solution $x(t)$ to $+\infty$ follows as in Theorem 7.6.

Theorem 7.5 implies that to every eigenvalue $\lambda(t)$ of $B(t) = [b_{ij}(t)]$ corresponds an eigenvalue λ' of $B = [b_{ij}]$ such that

$$|\lambda(t) - \lambda'| \leq (n + 2)M(t)r^{1/n}(t), \quad t \in R_+$$

where

$$M(t) = \max_{i, j=1, 2, \ldots, n} \{|b_{ij}(t)|, |b_{ij}|\}$$

$$r(t) = [1/(nM(t))] \sum_{i,j=1}^{n} |b_{ij}(t) - b_{ij}|$$

From the continuity of $B(t)$ and its convergence to B as $t \to +\infty$, we conclude that

$$\sup_{t \in R_+} M(t) = M < +\infty, \quad \lim_{t \to \infty} r(t) = 0$$

It follows that, given a constant $\epsilon > 0$, there exists a constant $T(\epsilon) > 0$ such that

$$|\lambda(t) - \lambda'| < \epsilon, \quad t \in [T(\epsilon), \infty)$$

If $\lambda(t) = 2q(t)$, then there exists an eigenvalue $2\tilde{q}$ of B such that

$$2|q(t) - \tilde{q}| < \epsilon, \quad t \in [T(\epsilon), \infty) \tag{7.20}$$

It should be noted that the eigenvalues λ', $\tilde{q}$ above actually depend on t.

Since B is positive definite, we may choose ϵ so that $0 < \epsilon < 2q$, where $2q$ is the smallest eigenvalue of B. Then (7.20) implies

$$2q(t) > 2\tilde{q} - \epsilon \geq 2q - \epsilon > 0, \quad t \in [T(\epsilon), \infty) \tag{7.21}$$

Thus, (7.19) and (7.21) give, for $\epsilon_1 = \epsilon/2$,

$$(d/dt)\|x(t)\|_V^2 \leq -2(q - \epsilon_1 - p(t))\|x(t)\|^2, \quad t \in [T(\epsilon), \infty) \tag{7.22}$$

Choosing $\epsilon_1 < q - p$ and $T = T(\epsilon)$ large enough, taking into consideration the number T in the statement of the theorem, we get

$$(d/dt)\|x(t)\|_V^2 \leq -2(q - p - \epsilon_1)\|x(t)\|^2 \tag{7.23}$$

$$\leq -2Q\|x(t)\|_V^2, \quad t \in [T, \infty)$$

where $Q = (q - p - \epsilon_1)/\mu$. Here μ is the largest eigenvalue of V. Applying Lemma 4.11, we obtain

$$\|x(t)\|_V \leq e^{-Q(t-T)}\|x(T)\|_V, \quad t \in [T, \infty) \tag{7.24}$$

Similarly, (7.19) leads to

$$\|x(t)\|_V \leq \exp\left\{\int_0^t [p(s) - q(s)]ds\right\}\|x_0\|_V \leq N_1\|x_0\|_V, \quad t \in [0, T] \tag{7.25}$$

ERRATA

Page	Line	Correction
7	20	$\|Tx_n\|$
23	21	ϵ_1-balls
25	6	$M \subset C_n^1[a,b]$
30	23	$h \in X$
31	19	$+ D^{-1}[v - f(u)],$
33	22	$i = 1, 2, \ldots, n$
37	1	$S \subset R^3$
37	2	$S_1 \subset R^3$
42	20	$(n+1)!$
44	1	Then
46	6	$x(t)$
52	6	M_n
63	27	$= \mu^{-1} K e^{-(m+1)\alpha T}$
70	23	$u(s) X(t) X^{-1}(s) X(s) ds$
74	8	$x(t)$
85	14	$\subset M_1$
88	4, 15	$G \subset Q_{m_1}, \quad t_1 + q\beta$
97	13	$\nabla V(u)$
99	16	T-periodic
101	16	(B_4)
109	14	by
110	22	$\ldots$ of $VS^\alpha \ldots$
114	12	delete $+ \|f\|_1]$
115	15	$\|x\|_\infty \leq \alpha$
138	7, 8, 20	$\mu V x, \quad \mu_0 V x, \quad -\mu V x$
169	11	For any bounded open $\ldots$
169	14	with $\|x_1 - x_2\|_\infty < \delta(\epsilon)$
172	19, 21	$V u_m, \quad = W u_m$
173	28	VS^α
174	1	VS^α
176	21	< 1

where N_1 is an upper bound for

$$\exp\{\int_0^t [p(s) - q(s)]ds\}, \quad t \in [0, T]$$

Combining (7.24) and (7.25), we arrive at

$$\|x(t)\| \leq N\|x_0\|, \quad t \in [0, T], \tag{7.26}$$

$$\|x(t)\| \leq N e^{-Q(t-T)} \|x_0\|, \quad t \in [T, \infty) \tag{7.27}$$

for a new positive constant N. Here we have used the equivalence of the norms $\|\cdot\|$, $\|\cdot\|_V$. It follows that the zero solution of (S_F) is asymptotically stable.

3. Stability Regions

Assume that $A: R_+ \to M_n$, $F: R_+ \times R^n \to R^n$ are continuous. Assume further that there exists $\Omega \subset R^n$ and $D \subset \Omega$ such that every solution $x(t)$ of (S_F) with $x(0) = x_0 \in D$ remains in Ω for $t \geq 0$. Then it is possible to introduce a concept of stability for (S_F), according to which the initial conditions are restricted to the subset D of Ω. The reason for doing this is that the system (S_F) possesses stability properties that stem from conditions on $A(t)$, $F(t, x)$ holding only on a subset Ω of R^n. We now define the concept of a stability (or asymptotic stability) region (D, Ω). Similarly one defines regions (D, Ω) corresponding to other stability concepts.

Definition 7.9 Consider the system (S_F) with $A: R_+ \to M_n$, $F: R_+ \times R^n \to R^n$ continuous and such that $F(t, 0) = 0$, $t \in R_+$. Assume further the existence of two subsets D, Ω, of R^n, with $D \subset \Omega$, such that $0 \in D$ and every solution $x(t)$ of (S_F) with $x(0) = x_0 \in D$ remains in Ω as long as it exists. Then the pair (D, Ω) is called a "region of stability" (asymptotic stability) for (S_F) if the zero solution of (S_F) is stable (asymptotically stable) w.r.t. initial conditions $x(0) = x_0 \in D$.

It is easy now to establish results like Theorems 7.7, 7.8, but with assumptions taking into consideration the sets D, Ω. Thus we have

Theorem 7.10 Let the system (S_F) satisfy the conditions on A, F, D, Ω of Definition 7.9. Furthermore, let $p: R_+ \to R$ be continuous and such that

$$\langle F(t, x), Vx \rangle \leq p(t)\|x\|^2, \quad t \in R_+, \ x \in \Omega \tag{7.28}$$

where V is as in Theorem 7.8. Let q be the smallest eigenvalue of $(1/2)B$.

Then if
$$\sup_{t \in R_+} p(t) < q$$
the pair (D, Ω) is a region of asymptotic stability for the system (S_F).

The proof is left as an exercise.

Example 7.11 In this example we search for an asymptotic stability region for the system
$$x' = Ax + F(t, x) \qquad (S_1)$$

where
$$A = \begin{bmatrix} 0 & 1 \\ -1 & -1 \end{bmatrix}, \quad F(t, u) = \begin{bmatrix} 0 \\ -(1/5)u_1^2 \end{bmatrix}$$

with $u = (u_1, u_2)$. This system arises from the scalar equation
$$y'' + y' + y + (1/5)y^2 = 0$$

We first note that the equation
$$A^T V + VA = -2I$$

has the (unique) solution $V = \begin{bmatrix} 3 & 1 \\ 1 & 2 \end{bmatrix}$. We shall determine a region Ω and a constant $\beta > 0$ such that
$$(d/dt)\|x(t)\|_V^2 \leq -2\beta \|x(t)\|_V^2, \quad t \in R_+ \qquad (7.29)$$

for any solution $x(t)$ of (S) that lies in Ω.

To this end, we first observe that Lemma 7.4 implies
$$(d/dt)\|x(t)\|_V^2 = 2 < A(t)x(t) + F(t, x(t)), Vx(t) > \qquad (7.30)$$
$$= -2[x_1^2(t) + x_2^2(t)] - (2/5)x_1^2(t)[x_1(t) + 2x_2(t)]$$

Thus, (7.29) will hold if the solution $x(t)$ satisfies

$$x_1^2(t) + x_2^2(t) + (1/5)x_1^2(t)[x_1(t) + 2x_2(t)]$$

$$\geq \beta \|x(t)\|_V^2 = \beta[3x_1^2(t) + 2x_1(t)x_2(t) + 2x_2^2(t)] \tag{7.31}$$

To determine Ω, we consider first the vectors $x \in R^2$ such that

$$x_1^2 + x_2^2 + (1/5)x_1^2(x_1 + 2x_2) \geq \beta(3x_1^2 + 2x_1x_2 + 2x_2^2) \tag{7.32}$$

Since $4x_1^2 + 3x_2^2 \geq 3x_1^2 + 2x_1x_2 + 2x_2^2$, (7.32) is satisfied for all $x \in R^2$ with

$$x_1^2 + x_2^2 + (1/5)x_1^2(x_1 + 2x_2) \geq \beta(4x_1^2 + 3x_2^2) \tag{7.33}$$

or

$$x_1^2[1 - 4\beta + (1/5)(x_1 + 2x_2)] + (1 - 3\beta)x_2^2 \geq 0$$

It follows that if we take $\beta = 1/5$, (7.32) is satisfied for all $x \in \Omega$, where

$$\Omega = \{x \in R^2; x_1 + 2x_2 \geq -1\}$$

Now we find a set $D \subset \Omega$ such that whenever a solution $x(t)$ of (S_1) starts inside D, it remains in Ω and satisfies (7.29). In this particular example, we can take D to be inside a ball D_1 in the V-norm; that is,

$$D_1 = \{x \in R^2; \|x\|_V \leq \sqrt{\alpha}\}$$

for some $\alpha > 0$. In fact, we first notice that the set D_1 consists of the interior of the ellipse

$$3x_1^2 + 2x_1x_2 + 2x_2^2 = \alpha \tag{7.34}$$

along with the ellipse itself. We also notice that Ω consists of all $x \in R^2$ which lie above the line $x_1 + 2x_2 = -1$. Thus we can determine α such that D_1 has a tangent point with this line and lies in Ω. Taking $\alpha = 1/2$, we find that $(0, -1/2)$ is this tangent point. We let $D = \{x \in R^2; \|x\|_V < \sqrt{\mu}\}$, for some $\mu \in (0, \alpha)$, and we show that every solution of (S_1) starting inside D, remains in D (see figure next page).

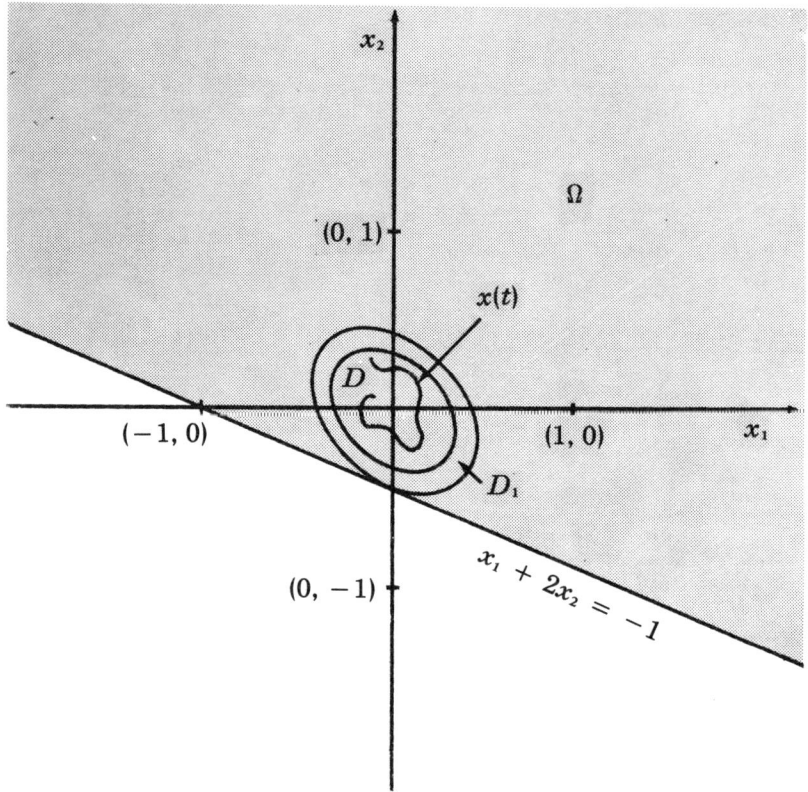

Let $x(t)$ be a solution of (S_1) starting at $x(0) = x_0$ with $x_0 \in D$. Then, as long as $x(t)$ lies in Ω, (7.29) is satisfied. Applying Lemma 4.11, we obtain

$$\|x(t)\|_V^2 \leq e^{-(2/5)t}\|x(0)\|_V^2 \leq \|x(0)\|_V^2 \tag{7.35}$$

If we assume that $x(t)$ leaves D_1 at some $t = t_0$, then there must be a neighborhood $[t_0, t_1)$ such that $x(t) \notin D$ and $x(t) \in D_1$ for all $t \in (t_0, t_1)$. Thus, again from 7.29, we obtain

$$(d/dt)\|x(t)\|_V^2 \big|_{t_0} \leq -(2/5)\|x(t_0)\|_V^2 = -(2/5)\mu < 0 \tag{7.36}$$

Since $(d/dt)\|x(t)\|_V^2$ is continuous, (7.36) shows that $\|x(t)\|_V$ is strictly decreasing in a right neighborhood of t_0. Thus, the solution $x(t)$ cannot penetrate the boundary of the V-ball D, because its V-norm does not increase to the right of the point t_0. It follows that (7.35) holds as long as $x(t)$ exists. This fact implies that $x(t)$ is continuable to $+\infty$. Moreover, (7.35) also implies that the pair (D, Ω) is a region of asymptotic stability for (S_1) (see 7.23-7.27). Actually, here (D, D) is an even better region of asymptotic stability.

4. Periodic Solutions

This section is devoted to the study of the system

$$x' = F(t, x) \qquad (E)$$

where $F(t + T, u) = F(t, u)$, $(t, u) \in R \times R^n$. Here T is a fixed positive constant. We are looking for T-periodic solutions of (E), that is, solutions $x(t)$ which exist on R and satisfy: $x(t + T) = x(t)$, $t \in R$.

Our first result is contained in Theorem 7.12.

Theorem 7.12 Let $F: R \times R^n \to R^n$ be continuous and such that $F(t + T, u) = F(t, u)$, $(t, u) \in R \times R^n$, where T is a positive constant. Moreover, let

$$< F(t, u_1) - F(t, u_2), u_1 - u_2 > \leq 0, \quad (t, u_1, u_2) \in [0, T] \times R^n \times R^n \qquad (7.37)$$

and

$$< F(t, u), u > < 0 \qquad (7.38)$$

for every $t \in [0, T]$ and every $u \in R^n$ with $\|u\| = r$, where r is a positive constant. Then (E) has at least one T-periodic solution.

Proof. We first note that Theorem 7.6 implies the existence and uniqueness w.r.t. initial conditions of solutions of (E) on R_+. Assume that $x(t)$ is a solution of (E) defined on $[0, T]$ and such that $x(0) = x(T)$. Then, by the T-periodicity of $F(t, u)$ w.r.t. to t, it is easy to define $x(t)$ on the entire real line so that it is T-periodic (see also Exercise 3.7). Thus, it suffices to show that (E) has a solution $x(t)$ defined on $[0, T]$ and such that $x(0) = x(T)$. To this end, let $B = \{u \in R^n; \|u\| \leq r\}$, and let $U: B \to X$ map every $u \in B$ to the value at T of the unique solution of the problem

$$x' = F(t, x), \quad x(0) = u, \quad t \in R_+$$

Then if $Ux_0 = x_0$, for some $x_0 \in B$, we have $x_0 = x(0) = x(T)$ for some solution $x(t)$ of (E), and the proof is complete. In order to apply the Brouwer theorem (Corollary 2.15), we have to show that $UB \subset B$ and that U is continuous. If $x(t)$, $y(t)$ are two solutions of (E) on $[0, T]$, then (7.12) implies

$$\|x(T) - y(T)\| \leq \|x(0) - y(0)\| \qquad (7.39)$$

which shows the continuity of U. Now let a solution $x(t)$ of (E) satisfy $\|x(0)\| \leq r$ and let

$$D = \{t \in [0, T]; \|x(s)\| \leq r \text{ for } s < t\}$$

We intend to prove that the number $c = \sup D$ is equal to T. In fact, let $c < T$ and suppose that $\|x(c)\| < r$. Then, since $x(t)$ is continuous, there exists an interval $[c, t_1) \subset [c, T)$ such that $\|x(s)\| < r$ there. However, this contradicts the definition of c. It follows that $\|x(c)\| = r$. Hence, by Lemma 7.4, we obtain

$$(d/dt)\|x(t)\|^2\big|_{t=c} = 2 < x'(c), x(c) >$$

$$= 2 < F(c, x(c)), x(c) >$$

$$< 0 \qquad (7.40)$$

Since $(d/dt)\|x(t)\|^2$ is continuous, there exists an interval $[c, t_2) \subset [c, T)$ such that $(d/dt)\|x(t)\|^2 < 0$ there. Integrating this last inequality once, we obtain $\|x(s)\| < \|x(c)\| = r$ for all s in this interval. This is a contradiction to the definition of c. It follows that $UB \subset B$ and that U has a fixed point in B. The proof is complete.

In our second result the operator U above is a contraction on R^n.

Theorem 7.13 Let $F: R \times R^n \to R^n$ be continuous and such that $F(t + T, u) = F(t, u)$, $(t, u) \in R \times R^n$, where T is a positive constant. Moreover, let M be a negative constant such that

$$< F(t, u_1) - F(t, u_2), u_1 - u_2 > \le M\|u_1 - u_2\|^2$$

for all $(t, u_1, u_2) \in [0, T] \times R^n \times R^n$. Then there exists a unique T-periodic solution of the system (E).

Proof. Continuation and uniqueness w.r.t. initial conditions follow from Theorem 7.6. It suffices to show that the operator U, defined in the proof of Theorem 7.12, has a unique fixed point in R^n. In fact, if $x(t), y(t)$ are two solutions of (E) on $[0, T]$, we have

$$(d/dt)[e^{-2Mt}\|x(t) - y(t)\|^2]$$

$$= -2Me^{-2Mt}\|x(t) - y(t)\|^2 + 2e^{-2Mt} < F(t, x(t)) - F(t, y(t)), x(t) - y(t) >$$

$$\le (-2Me^{-2Mt} + 2Me^{-2Mt})\|x(t) - y(t)\|^2$$

$$= 0$$

It follows, by integration, that

$$e^{-2Mt}\|x(t) - y(t)\|^2 \le \|x(0) - y(0)\|^2$$

or that
$$\|x(T) - y(T)\| \le e^{MT}\|x(0) - y(0)\|$$
Since $M < 0$, the operator U is a contraction mapping on R^n and, as such, it has a unique fixed point.

5. Boundary Value Problems on Infinite Intervals

Consider the problem

$$x' = A(t)x + F(t, x), \quad (S_F)$$

$$Ux = r \quad (B)$$

where U is a bounded linear operator on $C_n(R_+)$ and r is a fixed vector in R^n. We show here that it is possible to solve this problem under certain monotonicity assumptions on $A(t)$, $F(t, x)$, and for boundary conditions (B) which do not include periodicity conditions. This result is contained in the following theorem.

Theorem 7.14 Let $A: R_+ \to M_n$, $F: R_+ \times R^n \to R^n$ be continuous, $F(t, 0) = 0$, $t \in R_+$. Moreover, assume the following:

(i) for every $(t, u, \mu) \in R_+ \times R^n \times [0, 1]$ we have
$$< A(t)u + \mu F(t, u), u > \le 0;$$

(ii) $\|X(t)\| \le M$, $t \in R_+$, $M > 0$ constant, and there exist two functions $p, g: R_+ \to R$, continuous, and such that
$$\|X^{-1}(t)F(t, u)\| \le q(t)\|u\| + g(t), \quad (t, u) \in R_+ \times R^n$$
Here $X(t)$ denotes the fundamental matrix of the system
$$x' = A(t)x \quad (S)$$
with $X(0) = I$. Moreover,
$$\int_0^\infty q(t)dt < +\infty, \quad \int_0^\infty g(t)dt < +\infty$$

(iii) $U: C_n(R_+) \to R^n$ is a bounded linear operator such that $\|Uu\| \ge \phi(\|u(0)\|)$ for every $u \in C_n(R_+)$ with $\|u(t)\| \le \|u(0)\|$, $t \in R_+$. Here $\phi: R_+ \to R_+$ is continuous, strictly increasing and such that $\phi(0) = 0$.

(iv) $\widetilde{X}^{-1}$ exists, where $\widetilde{X}$ is defined in Chapter Six, Section 1.

Then the problem $((S_F), (B))$ has at least one solution.

Proof. Following the method developed in Chapter Six, we have to prove the existence of a fixed point for the operator V, where

$$(Vf)(t) = X(t)\widetilde{X}^{-1}[r - Up(\cdot, f)] + p(t, f)$$

$t \in R_+$. Let D_1, $V_1: D_1 \to D_1$ have the same meaning as in Theorem 6.14. There we indicated that the operator V_1 is compact and continuous in μ uniformly w.r.t. x in bounded subsets of D_1 (see Exercise 6.14). In order to apply the Leray-Schauder theorem (Theorem 2.16) we need to show that all possible solutions of the problem $x - \mu V_1 x = 0$ in D_1 lie inside a ball of D_1 which does not depend on $\mu \in [0, 1]$. In fact, let $x \in D_1$ solve $x - \mu_0 V_1 x = 0$ for some $\mu_0 \in [0, 1]$. Then the function $x(t)$ satisfies the integral equation

$$x(t) = \mu\{X(t)\widetilde{X}^{-1}[r - Up(\cdot, x)] + p(t, x)\} \tag{7.41}$$

for $\mu = \mu_0$ and for every $t \in [0, 1]$. It follows that $Ux = \mu_0 r$ and that

$$x'(t) = A(t)x(t) + \mu_0 F(t, x(t)), \quad t \in [0, 1] \tag{7.42}$$

Now we apply (7.6) for $V = I$ to get

$$(d/dt)\|x(t)\|^2 = 2 < A(t)x(t) + \mu_0 F(t, x(t)), x(t) > \leq 0, \quad t \in [0, 1]$$

This inequality immediately implies

$$\|x(t)\| \leq \|x(0)\|, \quad t \in [0, 1]$$

Consequently, by Hypothesis (iii), we obtain

$$\phi(\|x(t)\|) \leq \phi(\|x(0)\|) \leq \|Ux\| = \mu_0 \|r\| \leq \|r\|$$

Thus, we have a bound for $x(t)$: $\|x(t)\| \leq \phi^{-1}(\|r\|)$, $t \in [0, 1]$. The number $\phi^{-1}(\|r\|)$ is a uniform bound for all solutions of $x - \mu Sx$ in D_1, independently of $\mu \in [0, 1]$. Similarly, by induction, we obtain a sequence of functions $x_m(t)$, $t \in [0, m]$, $m = 1, 2, \ldots$, such that

$$\|x_m(t)\| \leq \phi^{-1}(\|r\|), \quad t \in [0, m]$$

and each $x_m(t)$ satisfies (7.41), for $\mu = 1$, on the interval $[0, m]$. The proof now follows as the proof of Theorem 6.14, and is therefore omitted.

Example 7.15 Consider the system

$$x' = \begin{bmatrix} -2 & 1 \\ 1 & -2 \end{bmatrix} x + F(t, x) \tag{7.43}$$

and the boundary condition

MONOTONICITY IN R^n AND DIFFERENTIAL SYSTEMS

$$Ux = x(0) + \int_0^\infty V(t)x(t)dt = r \qquad (7.44)$$

where $V: R_+ \to M_n$ is continuous and such that

$$\int_0^\infty \|V(t)\|dt < 1$$

Here we may take

$$F(t, x) = \begin{bmatrix} f_1(t)x_1 \\ f_2(t)x_2/(x_1^2 + x_2^2 + 1) \end{bmatrix}$$

where $f_1, f_2: R_+ \to R_-$ are continuous and such that

$$-\int_0^\infty \|X^{-1}(t)\|f_1(t)dt < +\infty, \quad -\int_0^\infty \|X^{-1}(t)\|f_2(t)dt < +\infty$$

Then $q(t) = -\|X^{-1}(t)\|f_1(t)$, $g(t) = -\|X^{-1}(t)\|f_2(t)$, and

$$\phi(\mu) = \left(1 - \int_0^\infty \|V(t)\|dt\right)\mu$$

In fact, if $u \in C_n(R_+)$ with $\|u(t)\| \leq \|u(0)\|$, we have

$$\|Uu\| = \left\|u(0) + \int_0^\infty V(t)u(t)dt\right\| \geq \|u(0)\| - \int_0^\infty \|V(t)\|\|u(t)\|dt$$

$$\geq \|u(0)\| - \int_0^\infty \|V(t)\|dt\|u(0)\| = \phi(\|u(0)\|)$$

The rest of the assumptions of Theorem 7.14 are also satisfied. Thus, the problem ((7.43), (7.44)) has at least one solution for every $r \in R^n$.

EXERCISES

7.1. Show that the solution $V \in M_n$ of (7.2), given by (7.3), is unique.

7.2. Show that the zero solution of the system

$$\begin{bmatrix} x_1 \\ x_2 \end{bmatrix}' = \begin{bmatrix} e^{-t} & 1 \\ -2 & -3 + [1/(t+1)] \end{bmatrix} \begin{bmatrix} x_1 \\ x_2 \end{bmatrix} + \begin{bmatrix} 0 \\ |\sin(tx_2)|(x_1 - x_2) \end{bmatrix}$$

is asymptotically stable by using Theorem 7.8.

7.3. Show that the first order scalar equation

$$x' + |\sin t|x + x^3 = \cos t$$

has at least one 2π-periodic solution.

7.4. Show that the system (in R^n):

$$x' = -Lx + F(t, x)$$

has a unique T-periodic solution if $L > 0$ is a constant, F is continuous, T-periodic in t and

$$\|F(t, u_1) - F(t, u_2)\| \le K\|u_1 - u_2\|$$

$(t, u_1, u_2) \in [0, T] \times R^n \times R^n$, where $K \in (0, L)$ is another constant.

7.5. Let $a: R_+ \to R$, $B: R_+ \to R_+$ be continuous. Show that all solutions to the problem

$$u' = a(t)u + b(t)|u|^{1/2}, \quad u(0) = u_0 \ge 0$$

are continuable to $+\infty$.

7.6. Let $F: [a, b] \times R^n \to R^n$ be continuous. Let $K \subset R^n$ be compact. Show that there exists a sequence of continuous functions $\{F_m\}_{m=1}^{\infty}$, $F_m: [a, b] \times K \to R^n$ such that

(i) $\lim\limits_{m \to \infty} \|F_m(t, u) - F(t, u)\| = 0$

uniformly on $[a, b] \times K$;

(ii) $\|F_m(t, u_1) - F_m(t, u_2)\| \le L_m\|u_1 - u_2\|$

for all $(t, u_1, u_2) \in [a, b] \times K \times K$, $m = 1, 2, \ldots$, where L_m is a positive constant.

7.7. Let $F: R_+ \times R^n \to R^n$ be T-periodic in its first variable and continuous. Show that the system (E) has at least one T-periodic solution under the mere assumption

$$< F(t, u), u > \le 0$$

for all $u \in R^n$ with $\|u\| = r$. Here r is a positive constant. This is a considerable improvement on Theorem 7.12. [Hint. Consider the systems

$$x' = F_m(t, x) - \delta_m x \qquad (*)$$

Here $\{F_m\}$ is the sequence in Exercise 7.6 defined on the closed ball with center at zero and radius $2r$, and $\{\delta_m\}_{m=1}^{\infty}$ is a decreasing sequence of positive constants such that $\delta_m \to 0$ as $m \to \infty$. Show that it is possible to choose a subsequence $\{\bar{F}_m\}$ of $\{F_m\}$ such that

$$< \bar{F}_m(t, u) - \delta_m u, u > \leq -\delta_m r^2/2$$

for $\|u\| = r$ and $m \geq 1$. Using the method of Theorem 7.12, obtain a T-periodic solution $x_m(t)$, $t \in [0, T]$, of (*) with F_m replaced by $\bar{F}_m$. Show that a subsequence $\{x_{m_k}(t)\}_{k=1}^{\infty}$ of $\{x_m(t)\}$ converges uniformly on $[0, T]$ to a T-periodic solution of the system (E).]

7.8. Using Exercise 7.7, show that the system

$$\begin{bmatrix} x_1 \\ x_2 \\ x_3 \end{bmatrix}' = - \begin{bmatrix} (\sin^2 t)(\|x\|^3 - 1) \\ (1 + x_1)^2 x_2 \\ x_2^4 x_3 \end{bmatrix}$$

has at least one 2π-periodic solution.

7.9. Assume that $A: R \times R^n \to R^n$, $B: R \times R^n \to R^n$ are continuous and T-periodic in their first variable t. Assume further that

$$< A(t, u), u > \leq -\lambda \|u\|^2, \quad \|B(t, u)\| \leq \mu$$

for every $(t, u) \in R_+ \times R^n$, where λ, μ are positive constants. Show that the system

$$x' = A(t, x) + B(t, x)$$

has at least one T-periodic solution. [Hint. Base your argument on the ball $S_r = \{u \in R^n; \|u\| \leq r\}$, where r is any number in $(\mu/\lambda, \infty)$. Use Exercise 7.7.]

7.10. Find a region of asymptotic stability (D, Ω) for the system (S_1) in the text, where

$$A = \begin{bmatrix} 0 & 1 \\ -1 & -1 \end{bmatrix}, \quad F(t, u) = \begin{bmatrix} 0 \\ -\lambda(t) u_1^2 \end{bmatrix}$$

Here $\lambda: R_+ \to R_+$ is continuous and such that $\lambda(t) \leq L$, $t \in R_+$, where L is a number in $[0, 1)$.

7.11. (Asymptotic equilibrium). Consider the system

$$x' = F(x) \tag{E$_1$}$$

with $F: R^n \to R^n$ continuous and such that

$$< F(u_1) - F(u_2),\ u_1 - u_2 >_V \le -\lambda \|u_1 - u_2\|^2$$

for every $(t, u_1, u_2) \in R_+ \times R^n \times R^n$. Here V is a positive definite matrix and λ is a positive constant. Show that (E$_1$) has a unique equilibrium solution, that is, a vector $\bar{x} \in R^n$ such that $F(\bar{x}) = 0$. [Hint. Pick an arbitrary solution $x(t)$, $t \in R_+$, of (E$_1$) and let $u(t) = x(t + h) - x(t)$, $t \in R_+$, where h is a fixed positive number. Show first that

$$(d/dt)\|u(t)\|_V^2 \le -2\lambda \|u(t)\|^2$$

and conclude, after appraising $\|u(t)\|$, that $x(t) \to \bar{x}$ as $t \to \infty$, where $\bar{x}$ is a finite vector. Then, from an inequality of the form

$$\|x'(t)\| \le M \exp\{-\lambda_1(t - t_0)\}\|x'(t_0)\|,\ \lambda_1,\ M \text{ positive constants}$$

conclude that $x'(t) \to 0$ as $t \to +\infty$.]

7.12. Show that the problem

$$x' = Ax + F(t, x),\ x(0) - Nx(\infty) = r$$

with

$$A = \begin{bmatrix} -1/2 & 0 \\ 0 & -1/3 \end{bmatrix},\quad F(t, u) = \begin{bmatrix} -e^{-t}u_1/(1 + u_1^2) \\ -e^{-2t}|\sin u_1|u_2 \end{bmatrix}$$

$$r = \begin{bmatrix} 1 \\ -1 \end{bmatrix}$$

has at least one solution if the norm of the matrix $N \in M_2$ is sufficiently small. Here

$$x(\infty) = \lim_{t \to \infty} x(t)$$

CHAPTER EIGHT

BOUNDED SOLUTIONS ON R; QUASILINEAR SYSTEMS; APPLICATIONS OF THE INVERSE FUNCTION THEOREM

In Chapter One we saw that if P is a projection matrix, then $I - P$ is also a projection matrix and $R^n = R_1 \oplus R_2$, where $R_1 = PR^n$ and $R_2 = (I - P)R^n$. In this chapter we show that the projection P may induce a "splitting" in the space of solutions of the system

$$x' = A(t)x \tag{S}$$

This means that the solutions of (S) with initial conditions $x_0 \neq 0$ in R_1 tend to zero as $t \to +\infty$, whereas the solutions with initial conditions x_0 such that $(I - P)x_0 \neq 0$ have norms that tend to $+\infty$ as $t \to +\infty$. A corresponding situation, with the roles of R_1, R_2 reversed, holds on R_-. This splitting occurs if the system (S) possesses a so-called "exponential dichotomy."

Exponential dichotomies for System (S) and their effect on the existence of bounded solutions on R of the system

$$x' = A(t)x + F(t, x) \tag{S_F}$$

are mainly the subject of Sections 1, 2.

Section 3 contains some results for so-called "quasilinear systems"

$$x' = A(t, x)x + F(t, x) \tag{S_Q}$$

where $A(t, u)$ is an $n \times n$ matrix. These results are intimately dependent upon the behavior of the linear systems

$$x' = A(t, f(t))x + F(t, f(t)) \qquad (S^f)$$

where the functions f belong to a suitable class.

In Section 4 we provide some interesting applications of the inverse function theorem to boundary value problems on finite and infinite intervals.

Several results concerning further asymptotic properties of nonlinear systems are included in the exercises at the end of this chapter.

1. Exponential Dichotomies

In what follows, P is a projection matrix in M_n. For convenience, we write $P_1 = P$ and $P_2 = I - P$. We then have $P_1 R^n = R_1$ and $P_2 R^n = R_2$. Here of course we have identified the projections P_1, P_2 with the corresponding bounded linear operators.

We now define the concept of the "angular distance" between the subspaces R_1 and R_2.

Definition 8.1 The "angular distance" $\alpha(R_1, R_2)$ between the subspaces $R_1 \neq \{0\}, R_2 \neq \{0\}$ is defined as follows:

$$\alpha(R_1, R_2) = \inf\{\|u_1 + u_2\|; u_k \in R_k, \|u_k\| = 1, k = 1, 2\}$$

The following lemma establishes a basic relationship between the norms of P_1, P_2 and $\alpha(R_1, R_2)$.

Lemma 8.2 Let $R^n = R_1 \oplus R_2$ and let P_1, P_2 be the projections of R_1, R_2, respectively. Then if $\|P_k\| > 0$, $k = 1, 2$, we have

$$1/\|P_k\| \leq \alpha(R_1, R_2) \leq 2/\|P_k\|, \quad k = 1, 2$$

Proof. Let $\mu > \alpha(R_1, R_2)$, where μ is a constant, and let $u_1 \in R_1, u_2 \in R_2$ be such that $\|u_k\| = 1$, $k = 1, 2$, and $\|u_1 + u_2\| < \mu$. Then if $u = u_1 + u_2$, we have $P_k u = u_k$, $k = 1, 2$, and

$$1 = \|u_k\| \leq \|P_k\| \|u\| < \mu \|P_k\|$$

Consequently, we find

$$1/\|P_k\| < \mu$$

which implies $1/\|P_k\| \leq \alpha(R_1, R_2)$.

Now let $u \in R^n$ be such that $P_k u \neq 0$, $k = 1, 2$. Then we have

$$\alpha(R_1, R_2) \leq \|[P_1 u/\|P_1 u\|] + [P_2 u/\|P_2 u\|]\|$$

$$= [1/\|P_1 u\|]\|P_1 u + [\|P_1 u\|/\|P_2 u\|] P_2 u\|$$

$$= [1/\|P_1 u\|]\|u + [(\|P_1 u\| - \|P_2 u\|)/\|P_2 u\|] P_2 u\|$$

$$\leq [1/\|P_1 u\|][\|u\| + [\|P_1 u + P_2 u\|/\|P_2 u\|]\|P_2 u\|]$$

$$\leq [1/\|P_1 u\|][\|u\| + \|u\|] = 2\|u\|/\|P_1 u\|$$

This implies

$$\sup\{[\|P_1 u\|/\|u\|]\alpha(R_1, R_2); P_k u \neq 0, k = 1, 2,\} \leq 2$$

or (see Exercise 8.1)

$$\|P_1\|\alpha(R_1, R_2) \leq 2$$

Similarly, $\|P_2\|\alpha(R_1, R_2) \leq 2$

We define below the "splitting" that was referred to in the introduction. The symbol $X(t)$ denotes again the fundamental matrix of (S) with $X(0) = I$.

Definition 8.3 Let $A: R \to M_n$ be continuous. We say that the system (S) possesses an "exponential splitting" if there exist two positive numbers H, m_0 and a projection matrix P with the following properties:

(i) Every solution $x(t) = X(t)x_0$ of (S) with $x_0 \in R_1$ satisfies

$$\|x(t)\| \leq H \exp\{-m_0(t - s)\}\|x(s)\|, \quad t \geq s \qquad (8.1)$$

(ii) Every solution $x(t) = X(t)x_0$ of (S) with $x_0 \in R_2$ satisfies

$$\|x(t)\| \leq H \exp\{-m_0(s - t)\}\|x(s)\|, \quad s \geq t \qquad (8.2)$$

(iii) $\|P_k\| \neq 0$, $k = 1, 2$, and if $P_1(t), P_2(t)$ denote the projections $X(t)P_1 X^{-1}(t)$, $X(t)P_2 X^{-1}(t)$, respectively, and $R_1(t) = P_1(t)R^n$, $R_2(t) = P_2(t)R^n$, then there exists a constant $\beta > 0$ such that

$$\alpha(R_1(t), R_2(t)) \geq \beta, \quad t \in R$$

In the following definition we introduce the concept of an exponential dichotomy. The existence of an exponential dichotomy for the system (S)(with $P_1 \neq 0, I$) is equivalent to the existence of an exponential splitting.

This is proved in Theorem 8.5.

Definition 8.4 Let $A: R \to M_n$ be continuous. We say that the system (S) possesses an "exponential dichotomy" if there exist two positive constants H_1, m_0 and a projection matrix P with the following properties:

$$\|X(t)P_1X^{-1}(s)\| \leq H_1 \exp\{-m_0(t-s)\}, \quad t \geq s$$

$$\|X(t)P_2X^{-1}(s)\| \leq H_1 \exp\{-m_0(s-t)\}, \quad s \geq t \tag{8.3}$$

Theorem 8.5 The system (S) possesses an exponential dichotomy with $P_1 \neq 0, I$ if and only if (S) possesses an exponential splitting.

Proof. Assume that the system (S) possesses an exponential dichtomy with $P_1 \neq 0, I$ and constants H_1, m_0 as in Definition 8.4. Let $x(t)$ be a solution of (S) with $x(0) = x_0 = P_1x_0 \in R_1$. Then we have $x(s) = X(s)x_0$ and $x_0 = X^{-1}(s)x(s)$. Thus,

$$\|x(t)\| = \|X(t)P_1x_0\| = \|X(t)P_1X^{-1}(s)x(s)\|$$

$$\leq H_1 \exp\{-m_0(t-s)\}\|x(s)\|, \quad t \geq s$$

The inequality (8.2) is proved in a similar way.

To show that $\alpha(R_1(t), R_2(t))$ has a positive lower bound, it suffices to observe, by virtue of Lemma 8.2, that $\|P_k(t)\| \leq H_1, t \in R$.

Conversely, assume that the system (S) possesses an exponential splitting with H, m_0 as in Definition 8.3. Let $x(t)$ be a solution of (S) with $x(0) = P_1X^{-1}(s)u$, for fixed $s \in R$ and $u \in R^n$. Then from Property (8.1) we obtain

$$\|x(t)\| = \|X(t)P_1X^{-1}(s)u\|$$

$$\leq H \exp\{-m_0(t-s)\}\|x(s)\|$$

$$= H \exp\{-m_0(t-s)\}\|X(s)P_1X^{-1}(s)u\|$$

$$\leq MH \exp\{-m_0(t-s)\}\|u\|$$

where the constant M is an upper bound for $\|P_1(t)\|$.

Here we have used the fact that the existence of a positive lower bound for $\alpha(R_1(t), R_2(t))$ is equivalent to the boundedness of the projections $P_1(t), P_2(t)$ on R.

The second inequality in (8.3) is proved in a similar way.

Now assume that (S) possesses an exponential splitting. It is easy to see that every solution $x(t)$ of (S) with $x(0) = P_1x(0)$ satisfies

$$\lim_{t \to \infty} \|x(t)\| = 0$$

Let $x(t)$ be a solution of (S) such that $x(0) \in R_2$. Then from Definition 8.3,

(ii) we obtain

$$\|x(s)\| \geq (1/H) \exp\{m_0(s - t)\}\|x(t)\|, \quad s \geq t$$

which implies

$$\lim_{s \to \infty} \|x(s)\| = +\infty$$

If $x(t)$ is now any solution of (S) with $P_2 x(0) \neq 0$, then $x(0) = x_1(0) + x_2(0)$ with $x_1(0) \in R_1$ and $x_2(0) \in R_2$.

It follows that

$$x(t) = X(t)x(0) = X(t)x_1(0) + X(t)x_2(0) \equiv x_1(t) + x_2(t)$$

with $x_1(t)$, $x_2(t)$ solutions of (S). From the above considerations and

$$\|x(t)\| \geq \|x_2(t)\| - \|x_1(t)\|$$

it follows that $\|x(t)\| \to +\infty$ as $t \to \infty$. Consequently, R_1 is precisely the space of all initial conditions of solutions of (S) which are bounded on R_+. The situation is reversed on the interval R_-.

Example 8.6 The system

$$\begin{bmatrix} x_1 \\ x_2 \end{bmatrix} = \begin{bmatrix} -1 & 0 \\ 0 & 1 \end{bmatrix} \begin{bmatrix} x_1 \\ x_2 \end{bmatrix} \tag{8.4}$$

possesses an exponential dichotomy. In fact, here we have

$$X(t) = \begin{bmatrix} e^{-t} & 0 \\ 0 & e^{t} \end{bmatrix}$$

and general solution

$$x(t) = X(t)x(0) = \begin{bmatrix} e^{-t} x_1(0) \\ e^{t} x_2(0) \end{bmatrix}$$

$$= X(t)P_1 x(0) + X(t)P_2 x(0)$$

where
$$P_1 = \begin{bmatrix} 1 & 0 \\ 0 & 0 \end{bmatrix}$$

It is easily seen that

$$X(t)P_1 X^{-1}(s) = \begin{bmatrix} e^{-(t-s)} & 0 \\ 0 & 0 \end{bmatrix}, \quad t \geq s,$$

$$X(t)P_2 X^{-1}(s) = \begin{bmatrix} 0 & 0 \\ 0 & e^{-(s-t)} \end{bmatrix}, \quad s \geq t$$

2. Bounded Solutions on R

In the previous section we established that in the presence of an exponential splitting the system (S) can have only one bounded solution on R — the zero solution. It is easy to see that this situation prevails even in the case of an exponential dichotomy with $P_1 = I$. Thus, if (S) possesses either one of these properties, the system

$$x' = A(t)x + f(t) \tag{S_f}$$

can have at most one bounded solution on R. Here f is any continuous function on R.

The following theorem ensures the existence of a bounded solution on R of the system (S_f). This solution has some interesting stability properties. We need Definition 8.7.

Definition 8.7 The zero solution of the system (S_F) is negatively unstable if there exists a number $r > 0$ with the following property: every other solution $x(t)$ of (S_F) defined on an interval $(-\infty, a]$, for any number a, satisfies

$$\sup_{t \leq a} \|x(t)\| > r$$

Theorem 8.8 Consider the system (S_F) under the following assumptions:

(i) $A: R \to M_n$ is continuous and such that the system (S) possesses an exponential dichotomy given by (8.3).

(ii) $F: R \times R^n \to R^n$ is continuous and, for some constant $r > 0$, there exists a function $\beta: R \to R_+$, continuous and such that

$$\|F(t, u) - F(t, v)\| \leq \beta(t)\|u - v\|$$

for every $t \in R$ and $u, v \in S_r = \{x \in R^n; \|x\| \le r\}$.

(iii) $\varrho = \sup_{t \in R} H_1[\int_{-\infty}^{0} \exp\{m_0 s\} \beta(s + t)ds + \int_{0}^{\infty} \exp\{-m_0 s\} \beta(s + t)ds] < 1;$

(iv) $\sup_{t \in R} \| \int_{-\infty}^{t} X(t)P_1 X^{-1}(s)F(s, 0)ds - \int_{t}^{\infty} X(t)P_2 X^{-1}(s)F(s, 0)ds \| < r(1 - \varrho)/2.$

Then (1) there exists a unique solution $x(t)$, $t \in R$, of the system (S_F) such that $\|x\|_\infty \le r$.

(2) If $P = I$, $F(t, 0) = 0$, $t \in R$, and if (iii) holds without necessarily the second integral, then the zero solution of (S_F) is negatively unstable.

(3) Let $P = I$ and let (iii) hold without necessarly the second integral. Let

$$\lambda = \lim_{t \to \infty} \sup (1/t) \int_0^t \beta(s)ds < m_0/H_1 \tag{8.5}$$

Then there exists a constant $\delta > 0$ with the following property: if $x(t)$ is the solution in Conclusion (1) and $y(t)$, $t \in [0, T)$, $0 < T < +\infty$, is another solution of (S_F) with

$$\|x(0) - y(0)\| \le \delta \tag{8.6}$$

then

$$\|y(t)\| \le r, \quad t \in [0, \infty) \tag{8.7}$$

and

$$\lim_{t \to \infty} \|x(t) - y(t)\| = 0 \tag{8.8}$$

Moreover, if $F(t, 0) = 0$, $t \in R^n$, then the zero solution of (S_F) is asymptotically stable.

Proof. (1) We consider the operator U defined as follows:

$$(Uf)(t) = \int_{-\infty}^{t} X(t)P_1 X^{-1}(s)F(s, f(s))ds - \int_{t}^{\infty} X(t)P_2 X^{-1}(s)F(s, f(s))ds \tag{8.9}$$

This operator satisfies $US^r \subset S^r$, where

$$S^r = \{f \in C_n(R); \|f\|_\infty \le r\}$$

Actually,

$$\|Uf\|_\infty \le r(1 + \varrho)/2 < r \tag{8.10}$$

Also, $\|Uf_1 - Uf_2\|_\infty \le \varrho\|f_1 - f_2\|$, $f_1, f_2 \in S^r$. (See Exercise 8.3)

By the Banach contraction principle, U has a unique fixed point $x(t)$ in the ball S^r. It is easy to see that the function $x(t)$ is a solution to the system (S_F) on R.

If $y(t)$, $t \in R$, is another solution of the system (S_F) with $\|y\|_\infty \le r$, we let $(Uy)(t) = g(t)$. Then the function $g(t)$ satisfies the equation

$$x' = A(t)x + F(t, y(t)) \tag{8.11}$$

However, by the dicussion at the beginning of this section, (8.3) implies that (8.11) can only have one bounded solution on R. Hence, $g(t) \equiv y(t)$, $t \in R$, and y is a fixed point of U in S^r. This says that $y(t) \equiv x(t)$, $t \in R$. Thus, $x(t)$ is unique.

(2) Now let $P = I$, $F(t, 0) = 0$, $t \in R$, and let (iii) hold without necessarily the second integral. Then, by what has been show above, zero is the only solution of (S_F) in the ball S^r. Assume that $y(t)$, $t \in (-\infty, a]$ is some solution of (S_F) such that

$$\sup_{t \le a}\|y(t)\| \le r$$

Then it is easy to see, as above, that the operator U^a with

$$(U^a f)(t) = \int_{-\infty}^{t} X(t)X^{-1}(s)F(s, f(s))ds$$

has a unique fixed point $x(t)$ in the ball

$$S_a^r = \{f \in C(-\infty, a]; \|f\|_\infty \le r\}$$

We must have $x(t) \equiv 0$, $t \in R_+$. The function $y(t)$ satisfies the equation

$$y(t) = X(t)[X^{-1}(t_0)y(t_0) + \int_{t_0}^{t} X^{-1}(s)F(s, y(s))ds] \tag{8.12}$$

for any $t_0, t \in (-\infty, a]$ with $t_0 \le t$. We fix t in (8.12) and take the limit of the right hand side as $t_0 \to -\infty$. This limit exists as a finite vector because

$$\|X(t)X^{-1}(t_0)y(t_0)\| \le rH \exp\{-m_0(t - t_0)\}, \quad t \ge t_0$$

We find
$$y(t) = X(t) \int_{-\infty}^{t} X^{-1}(s)F(s, y(s))ds$$

Consequently, y is a fixed point for the operator U^a in S_a^r. This says that $y(t) \equiv 0, t \in (-\infty, a]$.

It follows that every solution $y(t)$ of (S_F), defined on an interval $(-\infty, a]$, must be such that $\|y(t_m)\| > r$, for a sequence $\{t_m\}_{m=1}^{\infty}$ with $t_m \to -\infty$ as $m \to \infty$. Therefore, the zero solution of (S_F) is negatively unstable.

(3) Let the assumptions in (3) be satisfied (without necessarily $F(t, 0) \equiv 0$) and let the positive number $\delta < r(1 - \varrho)/2$ be such that

$$\|x(0) - y(0)\| < \delta$$

where $x(t)$ is the solution in (1) and $y(t)$ is another solution of (S_F) defined on the interval $[0, T)$, $0 < T < +\infty$. Then there exists a sufficiently small neighborhood $[0, T_1) \subset [0, T)$ such that $\|y(t)\| < r$ for $t \in [0, T_1)$. Here we have used the fact that $\|x\|_\infty < r(1 + \varrho)/2$ from (8.10). For such values of t, we obtain, from the variation of constants formula,

$$\|x(t) - y(t)\| \leq \|X(t)X^{-1}(0)\| \|x(0) - y(0)\|$$
$$+ \int_0^t \|X(t)X^{-1}(s)\| \beta(s) \|x(s) - y(s)\| ds$$

Using the dichotomy (8.3), we further obtain

$$\exp\{m_0 t\} \|x(t) - y(t)\| \leq H_1[\|x(0) - y(0)\| + \quad (8.13)$$
$$\int_0^t \exp\{m_0 s\} \beta(s) \|x(s) - y(s)\| ds]$$

An application of Gronwall's inequality to (8.13) yields

$$\|x(t) - y(t)\| \leq H_1 \|x(0) - y(0)\| \exp\{-m_0 t + H_1 \int_0^t \beta(s)ds\} \quad (8.14)$$

for $t \in [0, T_1)$. From the definition of λ in (3), we obtain that if $\epsilon > 0$ is such that $\lambda + 2\epsilon < m_0/H_1$, then there exists an interval $[t_0, \infty)$, $t_0 > 0$, such that

$$(1/t) \int_0^t \beta(s)ds < \lambda + \epsilon < (m_0/H_1) - \epsilon$$

for every $t \geq t_0$. This immediately implies

$$H_1 \int_0^t \beta(s)ds - m_0 t < -\epsilon H_1 t, \quad t \geq t_0$$

Thus, we may choose

$$\delta < \min\{r(1 - \varrho)/2H_1M, r(1 - \varrho)/2\}$$

where

$$M = \sup_{t \geq 0} \exp\{-m_0 t + H_1 \int_0^t \beta(s)ds\}$$

For such δ, (8.14) implies

$$\|x(t) - y(t)\| < r(1 - \varrho)/2 \tag{8.15}$$

for all $t \in [0, T)$. In fact, if (8.15) is assumed false, then, letting

$$T_2 = \inf\{t \in [0, T); \|x(t) - y(t)\| = r(1 - \varrho)/2\}$$

we obtain from (8.14) that

$$\|x(T_2) - y(T_2)\| < r(1 - \varrho)/2$$

that is, a contradiction. It follows that $y(t)$ is continuable to $+\infty$ and that $\|y\|_\infty < r$. If we further assume that $F(t, 0) = 0$, $t \in R_+$, then we must take $x(t) = 0$, $t \in R_+$. In this case (8.14) implies the asymptotic stability of the zero solution of (S_F) (see also Theorem 4.5).

In the next theorem the solution of Conclusion (1), Theorem 8.8, is shown to be T-periodic or almost periodic provided that the functions A, F have similar properties in the variable t. We need Definition 8.9.

Definition 8.9 The continuous function $F: R \times S \to R^n$ $(S \subset R^n)$ is said to be S-almost periodic if for every $\epsilon > 0$ there exists $l(\epsilon) > 0$ such that every interval of length $l(\epsilon)$ contains at least one number τ with

$$\|F(t + \tau, u) - F(t, u)\| < \epsilon, \quad (t, u) \in R \times S \tag{8.16}$$

In Chapter One we defined the concept of R^n-valued almost periodic functions. Naturally, a corresponding definition can be given for M_n-valued functions. This will be assumed in the following theorem.

Theorem 8.10 Let the assumptions (i)-(iv) of Theorem 8.8 be satisfied and let $x(t)$ be the solution in Conclusion (1) there. Then we have the following:

(i) if $A(t)$, $F(t, u)$ are T-periodic in t, then $x(t)$ is T-periodic;

(ii) let $P_1 = I$, and let (iii) of Theorem 8.8 hold without necessarily the second integral. Let $A(t)$, $F(t, u)$ be almost periodic, S_r-almost periodic, respectively. Then $x(t)$ is almost periodic.

Proof. (i) The function $z(t) = x(t + T)$, $t \in R$, is also a solution of (S_F) with $\|z\|_\infty \leq r$. In fact,

$$z'(t) = A(t + T)z(t) + F(t + T, z(t))$$
$$= A(t)z(t) + F(t, z(t))$$

Since $x(t)$ is unique in S^r, we must have $x(t) = x(t + T)$, $t \in R$. Hence, $x(t)$ is periodic with period T.

(ii) Given $\epsilon > 0$ we can find a number $l(\epsilon) > 0$ such that in each interval of length $l(\epsilon)$ there exists at least one number τ with

$$\|A(t + \tau) - A(t)\| < \epsilon, \quad \|F(t + \tau, u) - F(t, u)\| < \epsilon \quad (8.17)$$

for every $t \in R$, $u \in S_r$ (see also Exercise 1.19). We fix τ, ϵ and we let

$$\phi(t) = x(t + \tau) - x(t), \quad t \in R$$

Then $\phi(t)$ satisfies the equation

$$\phi' = A(t + \tau)\phi + Q(t) \quad (8.18)$$

where

$$Q(t) = [A(t + \tau) - A(t)]x(t)$$
$$+ F(t + \tau, x(t + \tau)) - F(t, x(t))$$

Using (8.17), we obtain

$$\|Q(t)\| \leq \epsilon r + \|F(t + \tau, x(t + \tau)) - F(t, x(t + t))\| + \|F(t, x(t + \tau)) - F(t, x(t))\|$$

$$\leq \epsilon r + \epsilon + \beta(t)\|\phi(t)\|$$

$$= \beta(t)\|\phi(t)\| + \epsilon(r + 1) \quad (8.19)$$

for all $t \in R$. It is easy to see now that

$$Y(t) = X(t + \tau)X^{-1}(\tau)$$

is the fundamental matrix of the system

$$y' = A(t + \tau)y \tag{8.20}$$

with $Y(0) = I$, and that this system possesses an exponential dichotomy given by (8.3), where X is replaced by Y, and $P_2 = 0$. Thus, for $\phi(t)$ we have

$$\phi(t) = \int_{-\infty}^{t} Y(t)Y^{-1}(s)Q(s)ds \tag{8.21}$$

for $t \in R$. In fact, the right hand side of (8.21) is a bounded solution of the system (8.18). However, this system has a unique bounded solution on R. Using the estimate of $\|Q(t)\|$ in (8.19), we obtain

$$\|\phi(t)\| \leq \int_{-\infty}^{t} \|Y(t)Y^{-1}(s)\| \|Q(s)\| ds$$

$$\leq H_1 \int_{-\infty}^{t} \exp\{-m_0(t-s)\}[\beta(s)\|\phi(s)\| + \epsilon(r+1)]ds$$

$$< \varrho\|\phi\|_\infty + (H_1/m_0)(r+1)\epsilon = \varrho\|\phi\|_\infty + \sigma\epsilon$$

where σ is another positive constant. It follows that

$$\|\phi(t)\| = \|x(t+\tau) - x(t)\| \leq [\sigma/(1-\varrho)]\epsilon$$

for every $t \in R$. We summarize the situation as follows: for every $\epsilon_1 > 0$ there exists $\bar{\ell}(\epsilon_1) = \ell((1-\varrho)\epsilon_1/\sigma)$ such that every interval of length $\bar{\ell}(\epsilon_1)$ contains at least one number τ such that

$$\|x(t+\tau) - x(t)\| \leq [\sigma/(1-\varrho)][(1-\varrho)\epsilon_1/\sigma] = \epsilon_1$$

for all $t \in R$. We have shown that $x(t)$ is almost periodic.

As it is expected, fixed points of the operator U in the proof of Theorem 8.8 can be obtained by use of the Schauder-Tychonov theorem (Theorem 2.12). Actually, it suffices to show the existence of a sequence $\{x_m(t)\}\{t\}_{m=1}^{\infty}$ such that $x_m: [-m, m] \to R^n$, $\|x_m\|_\infty \leq K$ ($K > 0$ constant), and

$$x_m = \int_{-m}^{t} X(t)P_1 X^{-1}(s)F(s, x_m(s))ds$$
$$- \int_{t}^{m} X(t)P_2 X^{-1}(s)F(s, x_m(s))ds \tag{E_m}$$

for every $m = 1, 2, \ldots, t \in [-m, m]$. Since each such function x_m satisfies the system (S_F) on $[-m, m]$, the existence of a solution $x(t)$ of (S_F) on R will follow from Theorem 3.9.

This process is followed in Theorem 8.11.

Theorem 8.11 Assume that $A: R \to M_n$ is continuous and such that the system (S) possesses an exponential dichotomy given by (8.3). Furthermore, assume that $F: R \times R^n \to R^n$ is continuous and such that

$$\liminf_{m \to \infty} 1/m \sup_{\substack{t \in R \\ \|u\| \leq m}} \{\|F(t, u)\|\} = 0 \tag{8.22}$$

Then System (S_F) has at least one bounded solution $x(t)$ on R.

Proof. We consider first the functional $q: C_n(R) \to R_+$ defined as follows:

$$q(f) = (2H_1/m_0)\|F(\cdot, f(\cdot))\|_\infty$$

We show that there exists $r > 0$ such that $q(S^r) \subset [0, r]$, where

$$S^r = \{f \in C_n(R); \|f\|_\infty \leq r\}$$

In fact, assume that the contrary is true and let $\{n_k\}_{k=1}^\infty$ be a sequence of positive integers such that

$$\lim_{k \to \infty} (1/n_k) \sup_{\substack{t \in R \\ \|u\| \leq n_k}} \{\|F(t, u)\|\} = 0 \text{ as } k \to \infty \tag{8.23}$$

This is possible by virtue of (8.22). Then we have that $q(S^{n_k}) \not\subset [0, n_k]$, $k = 1, 2, \ldots$. Consequently, each ball S^{n_k} contains at least one function f_{n_k} such that

$$q(f_{n_k}) = (2H_1/m_0)\|F(\cdot, f_{n_k}(\cdot))\|_\infty > n_k \tag{8.24}$$

This implies that

$$1 < (2H_1/m_0)\|F(\cdot, f_{n_k}(\cdot))\|_\infty / n_k \leq [(2H_1/m_0) \sup_{\substack{t \in R \\ \|u\| \leq n_k}} \{\|F(t, u)\|\}]/n_k$$

contradicting (8.22). Let $r > 0$ be such that $q(S^r) \subset [0, r]$ and let $U_1: C_n[-1, 1] \to C_n[-1, 1]$ be defined as follows:

$$(U_1 f)(t) = \int_{-1}^{t} X(t) P_1 X^{-1}(s) F(s, f(s)) ds - \int_{t}^{1} X(t) P_2 X^{-1}(s) F(s, f(s)) ds$$

Then $U_1 S_1^r \subset S_1^r$, where

$$S_1^r = \{f \in C_n[-1, 1]; \|f\|_\infty \leq r\}$$

In fact, given $f \in S_1^r$, let

$$\bar{f}(t) = \begin{cases} f(1), & t \geq 1 \\ f(t), & -1 \leq t \leq 1 \\ f(-1), & t \leq -1 \end{cases}$$

Then we have, for $t \in [-1, 1]$,

$$\|(U_1 f)(t)\| \le \int_{-1}^{t} \|X(t)P_1 X^{-1}(s)\| \|F(s, \bar{f}(s))\| ds$$
$$+ \int_{t}^{1} \|X(t)P_2 X^{-1}(s)\| \|F(s, \bar{f}(s))\| ds$$
$$\le H_1 \left[\int_{-1}^{t} \exp\{-m_0(t-s)\} ds + \int_{t}^{1} \exp\{-m_0(s-t)\} ds \right]$$
$$\times \|F(\cdot, \bar{f}(\cdot))\|_{\infty}$$
$$= H_1 \left[\int_{-t-1}^{0} \exp\{m_0 s\} ds + \int_{0}^{1-t} \exp\{-m_0 s\} ds \right] \|F(\cdot, \bar{f}(\cdot))\|_{\infty}$$
$$\le H_1 \left[\int_{-\infty}^{0} \exp\{m_0 s\} ds + \int_{0}^{\infty} \exp\{-m_0 s\} ds \right] \|F(\cdot, \bar{f}(\cdot))\|_{\infty}$$
$$= (2H_1/m_0) \|F(\cdot, \bar{f}(\cdot))\|_{\infty} = q(\bar{f}) \le r$$

It is easy to see that the set $U_1 S_1^r$ is relatively compact in $C_n[-1, 1]$ and that U_1 is continuous on S_1^r. By the Schauder-Tychonov theorem, U_1 has at least one fixed point $x_1 \in S_1^r$. Similarly, we obtain that each operator U_m, $m = 2, \ldots$, defined on $C_n[-m, m]$ by the right hand side of (E_m), has a fixed point x_m, $t \in [-m, m]$, such that $\|x_m\|_{\infty} \le r$ and (E_m) is satisfied on $[-m, m]$. Theorem 3.9 implies now the existence of a solution $x(t)$, $t \in R$, of (S_F) such that $\|x\|_{\infty} \le r$.

3. Quasilinear Systems

The most important property of a quasilinear system

$$x' = A(t, x)x + F(t, x) \tag{S_Q}$$

with $A: J \times R^n \to M_n$, $F: J \times R^n \to R^n$ (J a real interval), is that the system

$$x' = A(t, f(t))x + F(t, f(t)) \tag{S^f}$$

is linear. Here f is any R^n-valued function on J. Systems of the type (S^f) have been extensively studied in preceding chapters. It is therefore natural to ask whether information about (S_Q) can be obtained by somehow exploiting the properties of the system (S^f), where f belongs to a certain class $A(J)$ of continuous functions on J.

It is shown here that some of the properties of the system (S^f) can be carried over to the system (S_Q) via fixed point theory. In fact, if U denotes the operator that maps the function $f \in A(J)$ into the (unique) solution $x_f \in A(J)$ of (S^f), then the fixed points of U are solutions in $A(J)$ of the system (S_Q).

This procedure is followed here to obtain some stability and periodicity properties of the system (S_Q).

It should be noted that the quasilinear systems constitute quite a large class. To show this, it suffices to observe that if $B: J \times R^n \to R^n$ is continuously differentiable w.r.t. its second variable, then there exists a matrix $A(t, x)$ such that

$$B(t, x) = A(t, x)x + B(t, 0), \quad (t, x) \in J \times R^n \tag{8.25}$$

This assertion follows from the next lemma.

Lemma 8.12 Let $F: D \to R^n$ be continuously differentiable on D, where D is an open, convex subset of R^n. Let $x_0, x_1 \in D$ be given. Then we have

$$F(x_1) - F(x_0) = \int_0^1 F_x(sx_1 + (1-s)x_0)ds(x_1 - x_0) \tag{8.26}$$

where $F_x(u)$ is the Jacobian matrix $[\partial F_i(u)/\partial x_j]$, $i, j = 1, 2, \ldots, n$, of F at u.

Proof. Consider the function

$$g(s) = F(sx_1 + (1-s)x_0), \quad 0 \le s \le 1$$

This function is well defined because the set D is convex. Using the chain rule for vector valued functions, we have

$$g'(s) = F_x(sx_1 + (1-s)x_0)(x_1 - x_0) \tag{8.27}$$

Integrating (8.27) from $s = 0$ to $s = 1$, and recalling that $g(0) = F(x_0)$, $g(1) = F(x_1)$, we get (8.26).

We notice that if the function B in (8.25) satisfies $B(t, 0) \equiv 0$, then the system

$$x' = B(t, x)$$

becomes

$$x' = A(t, x)x \tag{8.28}$$

Before we state and prove the main stability result of this section, it is convenient to establish some definitions. In what follows, $A: R_+ \times R^n \to M_n$ will be assumed continuous on its domain.

Definition 8.13 The zero solution of the system (8.28) is called "weakly stable" if for every $\epsilon > 0$ there exists $\delta(\epsilon) > 0$ with the following property: for every $\xi \in R^n$ with $\|\xi\| \le \delta(\epsilon)$ there exists at least one solution $x \in C_n(R_+)$ of (8.28) satisfying $x(0) = \xi$ and

$$\|x\|_\infty \le \epsilon$$

This definition coincides with the usual definition of Chapter Four if the solutions of (8.28) are unique w.r.t. initial conditions at 0.

Definition 8.14 The systems
$$x' = A(t, f(t))x \tag{E_f}$$
are called "iso-stable" if for every $\epsilon > 0$ there exists $\delta(\epsilon) > 0$ with the property: for every $f \in C_n(R_+)$ with $\|f\|_\infty \le \epsilon$ and every solution $y_f(t)$, $t \in R_+$, of (E_f) with $\|y_f(0)\| \le \delta(\epsilon)$, we have $\|y_f\|_\infty \le \epsilon$.

The iso-stability of (E_f) is of course guaranteed by
$$\lim_{t \to \infty} \sup \int_0^t \mu(A(s, f(s)))ds = K < +\infty \tag{8.29}$$

where $K = K(\epsilon)$ does not depend on the particular function $f \in C_n(R_+)$ with $\|f\| \le \epsilon$. Here μ is the measure of Definition 4.7. This fact follows from Theorem 4.12.

The following theorem provides conditions for the stability of the zero solution of (8.28).

Theorem 8.15 Let $A: R_+ \times R^n \to R^n$ be continuous and such that the systems (E_f) are iso-stable. Then the zero solution of the system (8.28) is weakly stable. If, moreover, the solutions of the system (8.28) are unique w.r.t. initial conditions at zero, then the zero solution of the system (8.28) is stable.

Proof. Let $X_f(t)$ be the fundamental matrix of (E_f) with $X_f(0) = I$. Fix $\epsilon > 0$ and let $\delta(\epsilon) > 0$ be such that $\|X_f(t)\xi\| \le \epsilon$ for every $t \in R_+$, $f \in C_n(R_+)$ with $\|f\|_\infty \le \epsilon$, and $\xi \in R^n$ with $\|\xi\| \le \delta(\epsilon)$. This is possible by virtue of the iso-stability of the systems (E_f). Now fix $\xi \in R^n$ with $\|\xi\| \le \delta(\epsilon)$ and consider the set S consisting of all functions $f \in C_n[0, 1]$ such that $f(0) = \xi$ and $\|f\|_\infty \le \epsilon$. Obviously, S is a closed, convex and bounded set in the Banach space $C_n[0, 1]$. We define the operator $T: S \to C_n[0,1]$ as follows: given a function $f \in S$, $y = Tf$ is the solution of the system (E_f) with $y(0) = \xi$, restricted on the interval $[0,1]$. From our assumption above we obtain that, since $\|\xi\| \le \delta(\epsilon)$, $\|y\|_\infty \le \epsilon$. It follows that $TS \subset S$. If p is given by

$$p = \sup_{\substack{t \in [0,1] \\ \|u\| \le \epsilon}} \|A(t, u)\|$$

then $f \in S$ and $y = Tf$ imply

$$\|y'\|_\infty \le \|A(\cdot, f(\cdot))\|_\infty \|y\|_\infty \le p\epsilon$$

This says that the set TS is equicontinuous. Since it is also uniformly bounded, it is relatively compact by virtue of Theorem 2.5. In order to apply the Schauder-Tychonov theorem, it remains to show that T is continuous on S. Let $\{f_m\}_{m=1}^\infty \subset S$, $f \in S$ be such that

$$\lim_{m \to \infty} \|f_m - f\|_\infty = 0$$

APPLICATIONS OF THE INVERSE FUNCTION THEOREM   159

Then, since $\{Tf_m\}_{m=1}^\infty$ is uniformly bounded and equicontinuous, there exists a subsequence $\{Tf_{m_k}\}_{k=1}^\infty$ such that $u_k = Tf_{m_k} \to y \in S$ uniformly as $k \to \infty$. Let

$$u(t) = \xi + \int_0^t A(s, f(s))y(s)ds$$

The sequence $\{u_k\}$ satisfies

$$u_k(t) = \xi + \int_0^t A(s, f_{m_k}(s))u_k(s)ds$$

Subtracting the last two equations we get

$$\|u_k - u\|_\infty \le \int_0^1 \|A(s, f_{m_k}(s))u_k(s) - A(s, f(s))y(s)\|ds$$

which implies $\|u_k - u\|_\infty \to 0$ as $k \to \infty$. Thus, $u(t) = y(t)$, $t \in [0, 1]$. It follows that $y(t)$ satisfies the system (E_f) and that $Tf_{m_k} \to Tf$ as $k \to \infty$ uniformly on $[0,1]$. Since we could have started with an arbitrary subsequence of $\{f_m\}$ instead of $\{f_m\}$ itself, we have actually shown that every subsequence of $\{Tf_m\}$ contains a subsequence converging uniformly to Tf on $[0, 1]$. This is equivalent to saying that $Tf_m \to Tf$ as $m \to \infty$ uniformly on $[0, 1]$. Proceeding similarly, we obtain, by induction, a sequence $\{x_m\}_{m=1}^\infty$ of functions with the property: $x_m \in C_n[0, m]$, $x_m(0) = \xi$, $\|x_m(t)\| \le \epsilon$ for $t \in [0, m]$, and each x_m satisfies the system (8.28) on the interval $[0, m]$. From Theorem 3.9 we obtain now the existence of a solution $x(t)$, $t \in R_+$, of (8.28) such that $x(0) = \xi$ and $\|x\|_\infty \le \epsilon$. This proves the weak stability of the zero solution of (8.28).

If, in addition, the solutions of (8.28) are unique w.r.t. initial conditions at zero, then the solution $x(t)$ above is the only solution of (8.28) with $x(0) = \xi$. This proves the stability of the zero solution of (8.28).

Example 8.16 To illustrate Theorem 8.15, consider the system (8.28) with

$$A(t, u) = \begin{bmatrix} \exp[-(t - u_1^2)] & u_2/(t + 1)^2 \\ -u_2/(t + 2)^2 & \exp(-t^2) \end{bmatrix}$$

for any $(t, u_1, u_2) \in R_+ \times R \times R$. We choose, for convenience, the norm

$$\|x\|_2 = \max\{|x_1|, |x_2|\}$$

Then, using Table 4.13, we have

$$\mu(A(t, u)) = \max\{\exp[-(t - u_1^2)] + |u_2|/(t + 1)^2,$$

$$|u_2|/(t+2)^2 + \exp(-t^2)\} = \exp[-(t-u_1^2)] + |u_2|/(t+1)^2$$

for all $t \geq 0$. Given $f \in C_2(R_+)$ with $\|f\|_\infty \leq \epsilon$, we obtain

$$\mu(A(t, f(t))) = \exp[-(t - f_1^2(t))] + |f_2(t)|/(t+1)^2$$

$$\leq \exp(\epsilon^2 - t) + \epsilon/(t+1)^2, \ t \geq 1$$

Thus, (8.29) is satisfied. Since $A(t, u)u$ satisfies a Lipschitz condition w.r.t. u on any compact subset of $R_+ \times R^n$, the zero solution of System (8.28) is stable.

Example 8.17 Similarly, the zero solution of the system $x' = B(t, x)$ with

$$B(t, u) = \begin{bmatrix} e^{-t}(u_1^2 + u_2) + e^{-2t}u_2^2 \\ te^{-t}u_1^2 + (1 + t^2)^{-1}u_2\sin(u_2^2) \end{bmatrix}$$

is stable. The proof is left as an exercise (see Exercise 8.4).

Other stability properties of the system (8.28) can be studied by means of the preceding method. Since the corresponding statements and methods of proof are very similar to our above considerations, they are omitted.

As the reader might have expected, it is sometimes more convenient to transfer the properties of the systems (S^f) to the system (S_Q) without the use of fixed point theory. In fact, it might be advisable to pick a local solution $x(t)$ of (S_Q) and apply the assumed uniform conditions on (S^f) to the system

$$u' = A(t, \tilde{x}(t))u + F(t, \tilde{x}(t)) \tag{8.30}$$

or the system

$$u' = A(t, \tilde{x}(t))u + F(t, u) \tag{8.31}$$

where $\tilde{x}(t)$ concides with $x(t)$ on some interval and is constant everywhere else. To illustrate this procedure, we extend below Theorem 4.14 to quasilinear systems.

Theorem 8.18 For the system (S_Q) assume the following:
(i) $A: R_+ \times R^n \to M_n$ is continuous;
(ii) $F: R_+ \times R^n \to R^n$ is continuous and such that

$$\|F(t, x)\| \leq \lambda \|x\|, \ (t, x) \in R_+ \times R^n$$

where λ is a positive constant;

APPLICATIONS OF THE INVERSE FUNCTION THEOREM 161

(iii) there exists a positive constant K with the property: for every $f \in C_n(R_+)$ there exists a fundamental matrix $X_f(t)$ of the system (E_f) such that

$$\int_0^t \|X_f(t)X_f^{-1}(s)\|ds \leq K, \quad t \in R_+$$

and $\lambda K < 1$.

Then the zero solution of the system (S_Q) is asymptotically stable.

Proof. Let $x_0 \in R^n$ be given and let $x(t)$ be a local solution of (S_Q) defined on the interval $[0, p)$, for some constant $p > 0$. Let q be a constant in $(0, p)$ and consider the function

$$\tilde{x}(t) = \begin{cases} x(t), & t \in [0, q] \\ x(q), & t \in [q, \infty) \end{cases}$$

Then $\tilde{x} \in C_n(R_+)$ and the function $x(t)$ satisfies the system (8.31) on the interval $[0, q]$. Now we can follow the steps of the proof of Theorem 4.14 to show that $x(t)$ satisfies the inequality

$$\|x(t)\| \leq (1 - \lambda K)^{-1}\|x(0)\| \tag{8.32}$$

on the interval $[0, q]$. Since $q \in [0, p)$ is arbitrary, and the right hand side of (8.32) does not depend on q, it follows that $x(t)$ satisfies (8.32) on the entire interval $[0, p)$. Thus, by Theorem 3.8, $x(t)$ is continuable to the point $t = p$. It follows that $x(t)$ is continuable to $t = +\infty$ and that (8.32) holds on R_+. This shows that the zero solution of the system (S_Q) is stable. The proof of the fact that the zero solution of (S_Q) is asymptotically stable follows again as in Theorem 4.14, because we have

$$\int_0^t \|X_x(s)X_x^{-1}(s)\|ds \leq K, \quad t \geq 0$$

for every solution $x(t)$, $t \in R_+$, of (S_Q).

The existence of T-periodic solutions of the system (S_Q) is the content of the following theorem.

Theorem 8.19 Assume that $A: R \times R^n \to M_n$ is a symmetric matrix, T-periodic in its first variable, and such that its largest eigenvalue $\lambda_M(t, u)$ is bounded above by a negative constant $-q$ for all $(t, u) \in R \times R^n$. Let $F: R \times R^n \to R^n$ be continuous, T-periodic in t, and such that

$$\liminf_{m \to \infty} \{(1/m) \int_0^T \sup_{\|u\| \leq m} \|F(t, u)\|dt\} = 0$$

Then the system (S_Q) has at least one T-periodic solution.

Proof. Assume for the moment that for every continuous T-periodic $f \in C_n(R)$ the system (S^f) has a unique T-periodic solution $x_f(t)$. Then, following the theory developed in Chapter Six, it is easy to see that

$$x_f(t) = X_f(t)[I - X_f(T)]^{-1} X_f(T) \int_0^T X_f^{-1}(s) F(s, f(s)) ds$$
$$+ \int_0^t X_f(t) X_f^{-1}(s) F(s, f(s)) ds$$

where $X_f(t)$ is the fundamental matrix of (E_f) with $X_f(0) = I$. Obviously, the fixed points of the operator $U: f \to x_f$ are T-periodic solutions of the system (S_Q). In order to apply the Schauder-Tychonov theorem, we first show that the system (E_f) has indeed a unique T-periodic solution for every T-periodic function f. Fix $f \in P_n(T)$ and let $x_f(t)$, $t \in R_+$, be a solution of the system (E_f). Then we have

$$(d/dt)[\exp\{2qt\} \| x_f(t) \|^2] = 2q \exp\{2q(t)\} \| x_f(t) \|^2$$
$$+ 2\exp\{2qt\} < A(t, f(t)) x_f(t), x_f(t) >$$
$$\leq 2q \exp\{2qt\} \| x_f(t) \|^2 - 2q \exp\{2qt\} \| x_f(t) \|^2$$
$$= 0 \tag{8.33}$$

Integrating (8.33) from s to $t \geq s$, we get

$$\exp\{2qt\} \| x_f(t) \|^2 \leq \exp\{2qs\} \| x_f(s) \|^2 \tag{8.34}$$

Since $x(t) = X_f(t) X_f^{-1}(s) x_f(s)$, (8.34) implies

$$\| X_f(t) X_f^{-1}(s) x_f(s) \|^2 \leq \exp\{-2q(t - s)\} \| x_f(s) \| \tag{8.35}$$

Since $x_f(s)$ is an arbitrary vector in R^n, (8.35) yields

$$\| X_f(t) X_f^{-1}(s) \| \leq \exp\{-q(t - s)\}, \ t \geq s \tag{8.36}$$

Letting $s = 0$ in (8.36), we obtain that $\| X_f(t) \| \to 0$ as $t \to \infty$. This says that every solution $x(t)$, $t \in R_+$, of the system (E_f) tends to zero as $t \to \infty$. It follows that the only possible periodic solution of (E_f) is the zero solution.

Now we employ Theorem 6.1 to conclude that for each $f \in P_n(T)$ the system (S^f) has a unique T-periodic solution.

We need to show that $[I - X_f(T)]^{-1}$ is bounded uniformly w.r.t. $f \in P_n(T)$. We are considering only $t \in [0, T]$, because, as it was established in Chapter

6, the existence of an $x \in C_n[0, T]$, which is T-periodic, is sufficient for the existence of a T-periodic solution of (S_ϱ). Since (8.36) implies

$$\|X_f(T)\| \le \exp\{-qT\},$$

we have

$$\|[I - X_f(T)]\xi\| \ge \|\xi\| - \|X_f(T)\|\|\xi\| \ge (1 - \exp\{-qT\})\|\xi\|$$

for every $\xi \in R^n$. Thus, we have shown that

$$\|[I - X_f(T)]^{-1}\| \le (1 - \exp\{-qT\})^{-1}$$

for every $f \in P_n(T)$. The rest of the proof follows now as in Theorem 8.11 in order to show that U has a fixed point in some closed ball of $P_n(T)$ with center at zero. It is therefore omitted.

Example 8.20 Theorem 8.19 is well illustrated by the following example.

Consider the system

$$x' = \begin{bmatrix} -p(t, x) & 0 & 0 \\ 0 & -q(t, x) & 0 \\ 0 & 0 & -r(t, x) \end{bmatrix} x + F(t, x)$$

where $p, q, r: R \times R^3 \to R$ are continuous, 2π-periodic in t and such that

$$p(t, u) \ge 1, \quad q(t, u) \ge 2, \quad r(t, u) \ge 3$$

for every $(t, u) \in [0, 2\pi] \times R^3$. Moreover, $F: R \times R^3 \to R^3$ is continuous, 2π-periodic in t and such that

$$\|F(t, u)\| \le \lambda \|u\|^\sigma + \mu, \quad t \in [0, \pi], u \in R^3$$

where λ, μ, σ are positive constants with $\sigma \in (0, 1)$. Then $\lambda_M(t, u) \le -1$ and the rest of the assumptions of Theorem 8.19 are satisfied.

4. Applications of the Inverse Function Theorem

Our applications of the inverse function theorem are concerned with boundary value problems

$$x' = A(t)x + F(t, x) \tag{S_f}$$

$$Ux = r \tag{B}$$

of the type considered in Chapter Six. We recall again that if the homogeneous problem ($F \equiv 0, r = 0$) has only the zero solution, then the problem ((S_F), (B)) is equivalent to the problem

$$x(t) = X(t)\widetilde{X}^{-1}[r - Up(\cdot, x)] + p(t, x) \qquad (8.37)$$

where $\widetilde{X}$ is the matrix whose columns are the values of U on the corresponding columns of $X(t)$ and

$$p(t, x) = \int_0^t X(t)X^{-1}(s)F(s, x(s))ds$$

Here, and in what follows, we assume for convenience that F is defined on all of R^n w.r.t. x, and that $F: [0, T] \times R^n \to R^n$ and $A: [0, T] \to M_n$ are continuous. Let x_0 be a fixed element of $C_n[0, T]$, and let

$$S^r = \{x \in C_n[0, T]; \|x - x_0\|_\infty < r\} \qquad (8.38)$$

where r is a positive constant. Then the operator $T_0: S^r \to C_n[0, T]$ given by

$$(T_0 f)(t) = F(t, f(t)) \qquad (8.39)$$

is continuous on S^r. This follows as in Example 2.27 because F is uniformly continuous on the set $[0, T] \times S$, where

$$S = \{u \in R^n; \|u\| < r + \|x_0\|_\infty\}$$

If F has a continuous Jacobian matrix $F_x(t, u)$ on $[0, T] \times S$, then it is also easy to see, again as in Example 2.27, that the operator T_0 is Frechet differentiable at x_0 with Frechet derivative $T_0'(x_0)$ satisfying

$$[T_0'(x_0)h](t) = F_x(t, x_0(t))h(t) \qquad (8.40)$$

for every $h \in C_n[0, T]$ and every $t \in [0, T]$.

We now state the main result on finite intervals.

Theorem 8.21 For the equation (8.37) assume the following:

(i) $A: [0,T] \to M_n$, $F: [0, T] \times R^n \to R^n$ are continuous.

(ii) The Jacobian matrix $F_x(t, u)$ is defined and continuous on the set $[0, T] \times R^n$.

(iii) Let $x_0 \in C_n[0, T]$ be fixed, and let $f_0 \in C_n[0, T]$ be given by

$$x_0(t) = f_0(t) - X(t)\widetilde{X}^{-1}Up(\cdot, x_0) + p(t, x_0) \qquad (8.41)$$

Assume that the equation

$$x(t) = f(t) - X(t)\widetilde{X}^{-1}Uq(\cdot, x_0, x) + q(t, x_0, x) \qquad (8.42)$$

has a unique solution $x \in C_n[0, T]$ for every $f \in C_n[0, T]$, where

$$q(t, x_0, x) = \int_0^t X(t)X^{-1}(s)F_x(s, x_0(s))x(s)ds \qquad (8.43)$$

Then there exist two constants $\alpha > 0$, $\beta > 0$ with the property: for every $f \in C_n[0, T]$ with $\|f - f_0\|_\infty \leq \beta$ there exists a unique solution x to the equation

$$x(t) = f(t) - X(t)\widetilde{X}^{-1}Up(\cdot, x) + p(t, x) \qquad (8.44)$$

with the property $\|x - x_0\|_\infty \leq \alpha$.

Proof. Consider the operator V defined as follows:

$$(Vx)(t) = x(t) + X(t)\widetilde{X}^{-1}Up(\cdot, x) - p(t, x) \qquad (8.45)$$

It is easy to see that V is continuous on $C_n[0, T]$, and that it is Frechet differentiable at x_0 (see discussion on (8.40) and Exercise 2.6) with derivative $V'(x_0)$ given by

$$[V'(x_0)h](t) = X(t)\widetilde{X}^{-1}Uq(\cdot, x_0, h) - q(t, x_0, h)$$

Now fix $f \in C_n[0, T]$ and consider the equation $V'(x_0)h = f$. Our assumption (iii) implies that h is the unique solution of the linear equation (8.42). Thus, $V'(x_0)$ is bounded, one-to-one, and onto. Let, for some constant $r > 0$,

$$D = \{u \in R^n; \|u\| < r + \|x_0\|_\infty\},$$

$$D_1 = \{x \in C_n[0, T]; \|x\|_\infty < r + \|x_0\|_\infty\}$$

Then given $\epsilon > 0$ there exists $\delta(\epsilon) > 0$ such that

$$\|F_x(\cdot, u_1) - F_x(\cdot, u_2)\|_\infty \leq \epsilon/2\mu \qquad (8.46)$$

for any $u_1, u_2 \in D$ with $\|u_1 - u_2\| \leq \delta(\epsilon)$. Here $\mu = \max\{\mu_1, \mu_2\}$ with

$$\mu_1 = \max_{t \in [0,T]} \int_0^t \|X(t)X^{-1}(s)\|ds,$$

$$\mu_2 = \|X\|_\infty \|\widetilde{X}^{-1}\| \|U\|\mu_1$$

($\|\cdot\|_\infty$ denotes sup-norm)

It follows that for every $x_1, x_2 \in D_1$ with $\|x_1 - x_2\|_\infty \leq \delta(\epsilon)$ we have

$$\|V'(x_1)h - V'(x_2)h\|_\infty \leq \epsilon \|h\|_\infty$$

for any $h \in C_n[0, T]$. Our assertion follows now from the inverse function theorem (Theorem 2.26).

The preceding theorem has the following important corollary.

Corollary 8.22 Let the assumptions of Theorem 8.21 be satisfied with $\|f_0\|_\infty \leq \beta$. Then there exist positive numbers α, μ such that for each $r \in R^n$ with $\|r\| \leq \mu$ there exists a unique solution $x(t)$ of the problem $((S_F), (B))$ satisfying $\|x - x_0\|_\infty \leq \alpha$.

Proof. Let $\epsilon > 0$ be such that $\|f_0\|_\infty + \epsilon < \beta$. Then since

$$\lim_{\|r\| \to 0} \sup_{t \in [0,T]} \|X(t)\widetilde{X}^{-1}r - f_0(t)\| = \|f_0\|_\infty$$

there exists $q(\epsilon) > 0$ such that

$$\sup_{t \in [0,T]} \|X(t)\widetilde{X}^{-1}r - f_0(t)\| \leq \|f_0\|_\infty + \epsilon < \beta$$

whenever $\|r\| \leq q(\epsilon) = \mu$. Thus, for every $r \in R^n$ with $\|r\| \leq \mu$, the equation (8.37) has a unique solution in the set $\{x \in C_n[0, T], \|x - x_0\|_\infty \leq \alpha\}$.

The next theorem solves the problem $((S_F), (B))$ on the interval R_+. The solutions actually belong to $C_n(R_+)$.

Theorem 8.23 For the problem $((S_F), (B))$ assume the following:

(i) $A: R_+ \to M_n$, $F: R_+ \times R^n \to R^n$ are continuous.

(ii) The fundamental matrix $X(t)$ ($X(0) = I$) satisfies

$$\sup_{t \in R_+} \int_0^t \|X(t)X^{-1}(s)\| ds < +\infty$$

Furthermore, $U: C_n(R_+) \to R^n$ is a bounded linear operator such that $\widetilde{X}^{-1}$ exists;

(iii) $F_x(t, u)$ exists and is continuous on $R_+ \times R^n$. Given a bounded set $M \subset R^n$, the sets $F(R_+ \times M)$, $F_x(R_+ \times M)$ are bounded and for every $\epsilon > 0$ there exists $\delta(\epsilon) > 0$ such that

$$\|F_x(t, u_1) - F_x(t, u_2)\| < \epsilon$$

for all $(t, u_1, u_2) \in R_+ \times M \times M$ such that $\|u_1 - u_2\| \leq \delta(\epsilon)$;

(iv) For every $f \in C_n(R_+)$ the equation (8.42) has a unique solution $x \in C_n(R_+)$, where the function q is given by (8.43).

Given $x_0 \in C_n(R_+)$, let f_0 be defined by (8.41). Then there exist two constants $\alpha > 0$, $\beta > 0$ with the property: for every $f \in C_n(R_+)$ with $\|f - f_0\|_\infty \leq \beta$ there exists a unique solution $x \in C_n(R_+)$ of the equation (8.37) such that $\|x - x_0\|_\infty \leq \alpha$. If, moreover, $\|f_0\|_\infty < \beta$, then the problem $((S_F), (B))$ has a unique solution for all $r \in R^n$ with $\|r\|$ sufficiently small.

Proof. The key element in the proof is the continuity of the operator V in (8.45). The Frechet differentiability of V follows from the Frechet differentiability of $Tx = F(\cdot, x(\cdot))$ and Exercise 2.6. The uniform continuity of $V'(x)$ on any bounded subset of $C_n(R_+)$ follows from an inequality like (8.46), as in Theorem 8.21. Now fix $x \in C_n(R_+)$ and let

$$G(t) = F(t, x(t) + h(t)) - F(t, x(t)), \quad t \in R_+$$

with x, $x + h \in S_\mu = \{u \in C_n(R_+); \|u\|_\infty < \mu\}$. Here μ is a positive constant.

Then, by the mean value theorem,

$$G_i(t) = F_i(t, x(t) + h(t)) - F_i(t, x(t))$$
$$= \langle \nabla F_i(t, x(t) + \theta_i(t)h(t)), h(t) \rangle, \quad i = 1, 2, \ldots, n$$

where the functions $\theta_i(t)$ lie in $(0, 1)$. Since $F_x(t, u)$ is bounded on $R_+ \times M$, for any $M \subset R^n$ bounded, it follows that $\|G\|_\infty \leq K\|h\|_\infty$, where K is a constant depending on μ. Thus, T is continuous at $x \in C_n(R_+)$. This implies easily the continuity of V at $x \in C_n(R_+)$.

The reader will not have any difficulty in applying the above considerations to equations

$$x(t) = f(t) + \int_{-\infty}^{t} X(t) P_1 X^{-1}(s) F(s, x(s)) ds$$
$$- \int_{t}^{\infty} X(t) P_2 X^{-1}(s) F(s, x(s)) ds \tag{8.47}$$

under suitable assumptions, where P_1 is a projection matrix in M_n. The conclusion in this case would be that the system (S_F) has bounded solutions on R if $f_0 = Vx_0$ has a sufficiently small norm $\|f_0\|_\infty$ (so that f can be taken identically equal to zero in (8.47)). Here V is the operator defined by (8.47) written as $Vx = f$.

Now we examine the problem

$$x' = F(t, x) \tag{E}$$

$$Ux = 0 \tag{B}$$

from a different point of view. We first notice that if U is a nonlinear operator, then we cannot in general reduce the problem ((E), (B)) to an integral equation of the type (8.37). We also observe that the problem ((E), (B)) is equivalent to the problem

$$Vx = [0, 0] \tag{8.48}$$

where $Vx = [Nx, Ux]$. Here

$$(Nx)(t) = x'(t) - F(t, x(t)) \tag{8.49}$$

Thus, solutions to the problem ((E), (B)) can actually be obtained from an application of the inverse function theorem to the operator V in (8.48). This is accomplished below for boundary conditions (B), where U has a Frechet derivative at any $x_0 \in C_n^1(R_+)$. Thus, we obtain solutions of ((E), (B)) in $C_n^1(R_+)$. Another theorem is given extending the above result to problems

$$x' = F(t, x) + G(t, x) \tag{8.50}$$

$$Ux = Wx \tag{8.51}$$

with no differentiability assumptions on the function G and the nonlinear operator W. Two interesting corollaries cover the case of perturbations depending on a small parameter $\epsilon > 0$. Extensions to problems on R can be similarly treated, and they are therefore omitted.

We let $C_\ell^1 = C_n^1(R_+) \cap C_n^\ell$. The space C_ℓ^1 is a closed subspace of $C_n^1(R_+)$. Thus, it is a Banach space with norm

$$\|f\|_1 = \|f\|_\infty + \|f'\|_\infty$$

We also note that $C_n(R_+) \times R^n$ (with addition and multiplication by real scalars defined in the obvious way) is a Banach space with norm

$$\|[f, r]\| = \|f\|_\infty + \|r\|$$

The following condition on F will be needed in the sequel.

Condition (F) (i) $F: R_+ \times R^n \to R^n$ is continuous and $F(R_+ \times M)$ is bounded for every bounded set $M \subset R^n$. Moreover, the Jacobian matrix $F_x(t, u)$ exists and is continuous on $R_+ \times R^n$.

(ii) For every bounded set $M \subset R^n$, $F_x(R_+ \times M)$ is bounded and for every

$\epsilon > 0$ there exists $\delta(\epsilon) > 0$ such that

$$\|F_x(t, u_1) - F_x(t, u_2)\| < \epsilon$$

$(t, u_1, u_2) \in R_+ \times M \times M$.

If Condition (F) holds, then the operator $N: C_n^1(R_+) \to C_n(R_+)$ satisfies

$$(N(x + h))(t) - (Nx)(t) = h'(t) - F(t, x(t) + h(t)) + F(t, x(t))$$

and it is easy to see that N is Frechet differentiable at each $x_0 \in C_n^1(R_+)$ with Frechet derivative $N'(x_0)$ satisfying

$$[N'(x_0)h](t) = h'(t) - F_x(t, x_0(t))h(t), \quad t \in R_+$$

Theorem 8.24 Assume that the function F satisfies condition (F) and that the operator $U: C_n^1(R_+) \to R^n$ is continuous and Frechet differentiable at every $x_0 \in C_n^1(R_+)$. For any open set $S \subset C_n^1(R_+)$ we assume the following: for every $\epsilon > 0$ there exists $\delta(\epsilon) > 0$ such that

$$\|[U'(x_1) - U'(x_2)]h\| \le \epsilon \|h\|_\infty$$

for every $x_1, x_2 \in S$, $h \in C_n^1(R_+)$.

Let $x_0 \in C_n^1(R_+)$ be given and assume that the linear problem

$$x' - F_x(t, x_0(t))x = 0 \quad (8.52)$$

$$U'(x_0)x = 0 \quad (8.53)$$

has only the zero solution in $C_n^1(R_+)$. Assume further that

$$\sup_{t \in R_+} \int_0^t \|X(t)X^{-1}(s)\| ds < +\infty \quad (8.54)$$

where $X(t)$ is the fundamental matrix of (8.52) with $X(0) = I$.

Let $x_0'(t) - F(t, x_0(t)) = f_0(t)$, $t \in R_+$, $Ux_0 = r_0$. Then there exist numbers $\alpha > 0$, $\beta > 0$ such that for every $[f, r] \in C_n(R_+) \times R^n$ with $\|[f - f_0, r - r_0]\| \le \beta$, there exists a unique solution $x \in C_n^1(R_+)$ of the problem

$$x' = F(t, x) + f(t), \quad (8.55)$$

$$Ux = r \quad (8.56)$$

such that $\|x\|_1 \le \alpha$. If, in addition, $\|[f_0, r_0]\| \le \beta$, then the problem ((E), (B)) has a unique solution $x(t)$ with $\|x\|_1 \leq \alpha$.

Proof. It is easy to see that the operator V is Frechet differentiable at every $x \in C_n^1(R_+)$. The Frechet derivative $V'(x)$ is given by

$$[V'(x)h](t) = [h'(t) - F_x(t, x(t))h(t), U'(x)h]$$

$$= [[N'(x)h](t), U'(x)h] \qquad (8.57)$$

for every $t \in R_+$, $h \in C_n^1(R_+)$. To show that V is continuous at $x \in C_n^1(R_+)$, we observe that

$$\|Vx - Vy\| = \|[Nx - Ny, Ux - Uy]\|$$

$$= \|Nx - Ny\|_\infty + \|Ux - Uy\|$$

for any $y \in C_n^1(R_+)$. Thus, it suffices to show that N is continuous at x. To see this, we observe first that

$$\|(Nx)(t) - (Ny)(t)\| \leq \|x'(t) - y'(t)\| + \|F(t, x(t)) - F(t, y(t))\|$$

$$\leq \|x' - y'\|_\infty + \|F(\cdot, x(\cdot)) - F(\cdot, y(\cdot))\|_\infty \quad (8.58)$$

for every $x, y \in S_r = \{u \in C_n^1, \|u - x_0\|_1 < r\}$. Here r is a positive constant. As in the proof of Theorem 8.23, we can show that, for some positive number K,

$$\|F(\cdot, x(\cdot)) - F(\cdot, y(\cdot))\|_\infty \leq K\|x - y\|_\infty$$

provided that $x, y \in S_r$.

Thus, given $\epsilon > 0$ there exists $\delta(\epsilon) > 0$ such that $\delta(\epsilon) < \min\{\epsilon/2, \epsilon/2K\}$ and

$$\|Nx - Ny\|_\infty < \|x - y\|_1 + K\|x - y\|_1$$

$$< \epsilon/2 + \epsilon/2 = \epsilon$$

for every $x, y \in S_r$ with $\|x - y\|_1 < \delta(\epsilon)$. It follows that N is continuous at any $x \in C_n^1(R_+)$. The uniform continuity (in x) of $V'(x)$ (see inequality preceeding Corollary 8.22) in any bounded open subset of $C_n^1(R_+)$ follows from that of N' and U'. $V'(x_0)$ is a bounded linear operator because V is continuous at x_0. $V'(x_0)$ is one-to-one and onto because of our assumptions on the problem ((8.52), (8.53)) in connection with the remark following Theorem 6.1. Indeed, since the problem ((8.52), (8.53)) has only the zero solution in $C_n^1(R_+)$, the problem

$$x' = F_x(t, x_0(t))x + f(t)$$

$$U'(x_0)x = r$$

has a unique solution for every $[f, r] \in C_n(R_+) \times R^n$ given by

$$x(t) = X(t)\widetilde{X}^{-1}[r - U'(x_0)q(\cdot, f)] + q(t, f)$$

where

$$q(t, f) = \int_0^t X(t)X^{-1}(s)f(s)ds$$

The inverse function theorem (Theorem 2.26) implies our first assertion. Our second assertion follows from the first as in Corollary 8.22, and is therefore omitted.

Now we examine the problem ((8.50), (8.51)), where no differentiability conditions on G, W are assumed. We assume, for convenience, that $x_0 \equiv 0$, $f_0 \equiv 0$, $F(t, 0) \equiv 0$.

Theorem 8.25 Let the assumptions of Theorem 8.24 be satisfied with $C_n(R_+)$, $C_n^1(R_+)$ replaced everywhere by C_n^ℓ, C_ℓ^1 respectively. Furthermore, assume the following:

(i) If

$$q(t) = \sup_{\|u\| \in S_\alpha} \|F(t, u)\|$$

where α is given in the conclusion of Theorem 8.24 and $S_\alpha = \{u \in R^n; \|u\| \leq \alpha\}$, then

$$\int_0^\infty q(t)dt < +\infty$$

(ii) $G: R_+ \times S_\alpha \to R^n$ is continuous, and for every $\epsilon > 0$ there exists $\delta(\epsilon) > 0$ such that

$$\|G(t, u) - G(t, v)\| < \epsilon, \quad t \in R_+ \tag{8.59}$$

for every $(u, v) \in S_\alpha \times S_\alpha$. Moreover, if

$$\sigma(t) = \sup_{u \in S_\alpha} \|G(t, u)\|$$

then $\|\sigma\|_\infty + \|Wu\| \leq \beta$ for every $u \in C_n(R_+)$ with $\|u\|_\infty \leq \alpha$, where W is defined below, and

$$\int_0^\infty \sigma(s)ds < +\infty$$

(iii) W is defined and continuous on the set $S^\alpha = \{u \in C_n(R_+); \|u\|_\infty \leq \alpha\}$ with values in R^n.

Then the problem ((8.50), (8.51)) has at least one solution $x \in C_\ell^1$.

Proof. Consider the operator V which assigns to each function $u \in S^\alpha$ the unique solution x_u of the problem

$$x' = F(t, x) + G(t, u(t)) \tag{8.60}$$

$$Ux = Wu \tag{8.61}$$

that belongs to the ball $\{x \in C_\ell^1; \|x\|_1 \leq \alpha\}$.

The existence of such a solution is guaranteed by Theorem 8.24 which still holds if $C_n(R_+)$, $C_n^1(R_+)$ are replaced by the spaces C_n^ℓ and C_ℓ^1, respectively. The convergence of x_u to a finite limit $x_u(\infty)$ follows from

$$x_u(t) = x_u(0) + \int_0^t F(s, x_u(s))ds + \int_0^t G(s, u(s))ds$$

which, taking limits as $t \to \infty$, implies

$$x_u(\infty) = x_u(0) + \int_0^\infty F(s, x_u(s))ds + \int_0^\infty G(s, u(s))ds$$

The limits on the right hand side exist by virtue of the integral assumptions on p, σ.

To show that V is continuous, let $\{u_m\}_{m=1}^\infty \subset S^\alpha$ be given with $\|u_m - u\|_\infty \to 0$ and let $Tu_m = x_m$, $m = 1, 2, \ldots$. Then we have

$$x_m'(t) = F(t, x_m(t)) + G(t, u_m(t))$$

$$Ux_m = Tu_m$$

Since

$$\|x_m(t) - x_m(t')\| \leq 2 \left| \int_t^{t'} q(s)ds \right| + 2 \left| \int_t^{t'} \sigma(s)ds \right|$$

for every $m = 1, 2, \ldots$, it follows that $\{x_m\}$ is equicontinuous on R_+. Since it is also equiconvergent (Exercise 2.5, (iii)), there exists a subsequence $\{x_m^1(t)\}_{m=1}^\infty$ of $\{x_m(t)\}$ such that

$$\|x_m^1 - y\|_\infty = \|Vu_m^1 - y\|_\infty \to 0 \text{ as } m \to \infty$$

where y is some function in C_n^ℓ. Letting

$$v_m(t) = (d/dt)x_m^1(t), \quad t \in R_+$$

we also obtain

$$\|v_m(t) - v_k(t)\| \le \|F(t, x_m^1(t)) - F(t, x_k^1(t))\|$$
$$+ \|G(t, u_m^1(t)) - G(t, u_k^1(t))\| \tag{8.62}$$

It is easy to see now that the boundedness of $F_x(t, u)$ on $R_+ \times S_\alpha$ and (8.59) imply that $\{v_m(t)\}$ is a Cauchy sequence of functions (see proof of Theorem 8.23) in $C_n(R_+)$. Thus, by a well-known theorem of advanced calculus, we have

$$\lim_{m \to \infty} v_m(t) = y'(t) \text{ uniformly on } R_+ \tag{8.63}$$

We let

$$v(t) = y(0) + \int_0^t F(s, y(s))ds + \int_0^t G(s, u(s))ds$$

and observe that

$$x_m^1(t) = x_m^1(0) + \int_0^t F(s, x_m^1(s))ds + \int_0^t G(s, u_m^1(s))ds$$

Subtracting these two equations, we eventually obtain

$$\|v(t) - x_m^1(t)\| \le \|y(0) - x_m^1(0)\| + \int_0^\infty \|F(s, x_m^1(s)) - F(s, y(s))\| ds$$
$$+ \int_0^\infty \|G(s, u_m^1(s)) - G(s, u(s))\| ds$$

Using Lebesgue's dominated convergence theorem, along with the integral conditions on p, σ, we obtain that

$$\lim_{m \to \infty} \|x_m^1 - v\|_\infty = 0$$

Thus, $v(t) = y(t)$, $t \in R_+$. This shows that $y(t)$ solves (8.60) on R_+ and belongs to C_n^ℓ. From (8.63) we also obtain that $y' \in C_n(R_+)$. Consequently, $y \in C_\ell^1$ and is the unique solution of the problem ((8.60), (8.61)) in C_ℓ^1. Since we could have started with any subsequence of $\{u_m\}$ instead of $\{u_m\}$ itself, we have actually proved the following statement: every subsequence $\{Vu'_m\}$ of $\{Vu_m\}$ contains a subsequence, say $\{Vu'_{m_k}\}$, which converges to $y(t)$ as $k \to \infty$ in the norm of C_ℓ^1. This proves the continuity of V. The relative compactness of TS^α in C_n^ℓ follows from the equicontinuity, the equiconvergence, and the boundedness ($TS^\alpha \subset S^\alpha$) of TS^α. The Schauder-Tychonov theorem implies now the existence of a solution $x(t)$ of the problem ((8.50), (8.51)) which belongs to C_ℓ^1.

Corollary 8.26 Assume that the hypotheses of Theorem 8.24 are satisfied and let η be a positive constant. Let $f: R_+ \times (0, \eta) \to R^n$ be continuous and such that

$$\lim_{\epsilon \to 0^+} \|f(\cdot, \epsilon) - f_0(\cdot)\|_\infty = 0$$

Let $r: (0, \eta) \to R^n$ satisfy

$$\lim_{\epsilon \to 0^+} r(\epsilon) = 0$$

Then there exists $\epsilon_0 > 0$ such that for each $\epsilon \in (0, \epsilon_0]$ the problem

$$x' = F(t, x) + f(t, \epsilon)$$

$$Ux = r(\epsilon)$$

has a unique solution x_ϵ such that $\|x_\epsilon\|_1 \leq \alpha$.

Proof. It suffices to observe that $\|[f(\cdot, \epsilon), r(\epsilon)]\| \leq \beta$ for sufficiently small ϵ.

Corollary 8.27 Assume that the hypotheses of Theorem 8.25 are satisfied for $G(t, x)$, Wx replaced by $G(t, x, \epsilon)$, $W(x, \epsilon)$, respectively, for every ϵ in the interval $(0, \eta)$, $\eta > 0$. Let

$$\lim_{\epsilon \to 0^+} \sup_{\|x\|_\infty \leq \alpha} \|W(x, \epsilon)\| = 0,$$

$$\lim_{\epsilon \to 0^+} \sup_{\substack{t \in R_+ \\ \|x\| \leq \alpha}} \|G(t, x, \epsilon)\| = 0$$

Then there exists $\epsilon_0 > 0$ such that for every $\epsilon \in (0, \epsilon_0]$ the problem

$$x' = F(t, x) + G(t, x, \epsilon),$$

$$Ux = W(x, \epsilon)$$

has at least one solution $x_\epsilon \in C_\ell^1$ such that $\|x_\epsilon\|_1 \leq \alpha$.

EXERCISES

8.1. Show that, in the setting of Lemma 8.2,

$$\sup_{\substack{\|u\|=1 \\ u \in R_\ell}} \|P_1 u\| \leq \sup_{\substack{\|u\|=1 \\ P_k u \neq 0, k=1,2}} \|P_1 u\|, \quad \ell = 1, 2$$

8.2. Let $A: R \to M_n$ be continuous and let the system (S) possess an exponential splitting. Show that if $x(t)$ is a solution of (S) with $x(0) \in R_1$, $x(0) \neq 0$, then $x(t) \notin R_2(t)$ for any $t \in R$.

8.3. Show that $US' \subset S'$ and $\|Uf_1 - Uf_2\| \leq \varrho$, for $f_1, f_2 \in S'$, in the proof of Theorem 8.8, (1).

8.4. Show the stability of the zero solution of the system $x' = B(t, x)$, where the vector $B(t, u)$ is given in Example 8.17.

8.5 Consider the quasilinear system

$$x' = A(t, x)x + F(t, x) \quad (S_Q)$$

where $A: R_+ \times R^n \to M_n$, $F: R_+ \times R^n \to R^n$ are continuous. Assume that there exist two functions $p, q: R_+ \to R_+$, continuous and such that

$$\|A(t, u)\| \leq p(t), \|F(t, u)\| \leq q(t), (t, u) \in R_+ \times R^n$$

Show that every local solution $x(t)$ of (S_Q) is continuable to $+\infty$.

8.6. Show that the scalar quasilinear equation

$$x' = (\cos^2 x - 2)x + (\sin t)x^{1/3} + \sin(2t)$$

has a 2π-periodic solution.

8.7. Suppose that $F: R \times R^n \to R^n$ is continuous. Let the constant $r > 0$ be such that

$$< F(t, u), u > \leq 0$$

for every $u \in R^n$ with $\|u\| = r$. Then the system (E) has at least one solution $x(t)$, $t \in R$, such that $\|x\|_\infty \leq r$. [Hint. Consider the systems

$$x' = F(t, x) - \epsilon x \quad (S_\epsilon)$$

where $\epsilon > 0$. Obtain solutions $x_\epsilon(t)$, $t \in R$, of (S_ϵ) such that $\|x_\epsilon\| \leq r$.

(See proof of Theorem 7.12).]

8.8. Consider the system

$$x' = A(t, x)x + F(t, x) \qquad (S_\varrho)$$

where A, F, p, q are as in Exercise 8.5. Assume further that

$$\int_0^\infty p(t)dt < +\infty, \quad \int_0^\infty q(t)dt < +\infty$$

Show that for every vector $\xi \in R^n$ there exists at least one solution $x(t)$ of the system (S_ϱ) which is defined for all large t and converges to ξ as $t \to \infty$.

8.9. Assume that $A: R_+ \to M_n$, $F: R_+ \times R^n \to R^n$, $f: R_+ \to R^n$ are continuous. Assume that the system (S) possesses a "dichotomy" given by (8.3), but for $t, s \geq 0$ and $m_0 = 0$. Let the function F satisfy

$$\|F(t, u_1) - F(t, u_2)\| \leq \lambda(t)\|u_1 - u_2\|, \quad \int_0^\infty \|F(t, 0)\|dt < +\infty$$

for every $u_1, u_2 \in R^n$, where $\lambda: R_+ \to R_+$ is continuous and such that

$$\int_0^\infty \lambda(t)dt < +\infty$$

Show that for every bounded solution $y(t)$, $t \in R_+$, of the system

$$x' = A(t)x + f(t) \qquad (S_f)$$

there exists a unique bounded solution $x(t)$, $t \in R_+$, of the system

$$x' = A(t)x + f(t) + F(t, x) \qquad (S_1)$$

such that the operator $T: x \to y$ is one-to-one, onto, and bicontinuous (T, T^{-1} are continuous). [Hint. Choose t_1 so that

$$H_1 \int_{t_1}^\infty \lambda(t)dt < +\infty$$

and find a unique fixed point $x(t)$ for the operator $V: C_n[t_1, \infty) \to C_n[t_1, \infty)$ with

$$(Vf)(t) = y(t) + \int_{t_1}^t X(t)P_1 X^{-1}(s)F(s, f(s))ds$$
$$- \int_t^\infty X(t)P_2 X^{-1}(s)F(s, f(s))ds$$

Show that $x(t)$ is continuable (in a unique way) to the point $t = 0$.

APPLICATIONS OF THE INVERSE FUNCTION THEOREM   177

Study the correspondence $T: x \to y$ with $y = x - Vx$, where $x \in C_n(R_+)$ is a solution of (S_1).]

8.10. Let $A: R_+ \times R^n \to M_n$ be continuous. Using the method of Theorem 8.15, show that the zero solution of the system

$$x' = A(t, x)x \qquad (8.28)$$

is uniformly asymptotically stable, provided that suitable stability properties are assumed for the systems: $x' = A(t, f(t))x$. In addition, examine the problem of strong stability of the system (8.28) via the same method.

8.11. Let $B: R_+ \times R^n \to R^n$, $F: R_+ \times R^n \to R^n$ be continuous with B continuously differentiable on $R_+ \times R^n$. Assume further that

$$\|B_x(t, u)\| \le p(t), \quad \|F(t, u)\| \le q(t)$$

for every $(t, u) \in R_+ \times R^n$, where $p, q: R_+ \to R_+$ are continuous and such that

$$\int_0^\infty p(t)dt < +\infty, \quad \int_0^\infty q(t)dt < +\infty$$

Show that for every solution $y(t)$, $t \in R_+$, of the system

$$y' = B(t, y)$$

there exists at least one solution $x(t)$, $t \in [t_1, \infty)$ (for some $t_1 \ge 0$), of the system

$$x' = B(t, x) + F(t, x)$$

such that $x(t) - y(t) \to 0$ as $t \to \infty$. [Hint. Consider the integral equation

$$u(t) = -\int_t^\infty A(t, u(s))ds - \int_t^\infty F_0(s, u(s))ds$$

where
$A(t, u) = B(t, u + y(t)) - B(t, y(t))$, $F_0(t, u) = F(t, u + y(t))$,
$u(t) \equiv x(t) - y(t)$.]

8.12. Consider the problem

$$\begin{bmatrix} x_1 \\ x_2 \end{bmatrix}' = \begin{bmatrix} -x_1^2 + x_2 \\ x_1 - x_2^2 \end{bmatrix} + \begin{bmatrix} f_1(t) \\ f_2(t) \end{bmatrix}$$

$$\begin{bmatrix} x_1(0) \\ x_2(0) \end{bmatrix} - \begin{bmatrix} x_1(2\pi) \\ x_2(2\pi) \end{bmatrix} = \begin{bmatrix} r_1 \\ r_2 \end{bmatrix} \tag{S_2}$$

Show that there exists a number $\delta > 0$ such that whenever $|r_1|, |r_2| < \delta$ and $|f_1|_\infty, |f_2|_\infty < \delta$ ($f_1, f_2 \in C_1[0, 2\pi]$), the problem ($S_2$) has at least one solution $x \in C_2[0, 2\pi]$.

8.13. Prove that the system

$$\begin{bmatrix} x_1 \\ x_2 \end{bmatrix}' = \begin{bmatrix} (|\sin x_1| - 3)x_1 + x_2 \\ x_1 + (\exp\{-|x_2|\} - 2)x_2 \end{bmatrix} + \begin{bmatrix} x_1^{1/3} \\ \sin^2 t \end{bmatrix}$$

has at least one 2π-periodic solution [Hint. Examine the eigenvalues of the matrix

$$\begin{bmatrix} |\sin f_1(t)| - 3 & 1 \\ 1 & \exp\{-|f_2(t)|\} - 2 \end{bmatrix}$$

for every $f \in C_2[0, 2\pi]$ which is 2π-periodic.]

8.14. Consider the quasilinear system

$$x' = A(t, x)x + f(t) \tag{8.28}$$

where $A: R \times R^n \to M_n$, $f: R \to R^n$ are continuous, T-periodic in t and such that

$$\|A(t, u) - A_1\| \le K$$

for every $(t, u) \in R \times R^n$, where $A_1 \in M_n$ is fixed. Show that if

$$x' = A_1 x, \quad x(0) = x(T)$$

has only the zero solution and K is sufficiently small, then the system (8.28) has at least one T-periodic solution.

8.15. Show that the function

$$x(t) = [1/(a - b)][(b - t)\int_a^t (s - a)f(s)ds + (t - a)\int_t^b (b - s)f(s)ds]$$

where $f \in C_1[a, b]$, is the unique solution to the scalar boundary value problem

$$x'' = f(t), \quad x(a) = x(b) = 0$$

Using the inverse function theorem, impose conditions on the function $F: [0, 1] \times R \to R$ that ensure the solvability of the problem

$$x'' = F(t, x), \quad x(0) = x(1) = 0$$

8.16. Complete the proof of Theorem 8.19.

BIBLIOGRAPHY

1. Bellman, R., *Introduction to Matrix Analysis*, McGraw-Hill, New York, 1960.
2. Bellman, R., *Stability Theory of Differential Equations*, Dover, New York, 1969.
3. Bernfeld, S. R., and V. Lakshmikantham, *An Introduction to Nonlinear Boundary Value Problems*, Academic Press, New York, 1974.
4. Besicovitch, A. S., *Almost Periodic Functions*, Dover, New York, 1954.
5. Brauer, F., and J. A. Nohel, *The Qualitative Theory of Ordinary Differential Equations*, W. A. Benjamin, New York, 1968.
6. Cesari, L. *Asymptotic Behaviour and Stability Problems in Differential Equations*, Springer-Verlag, New York, 1963.
7. Coddington, E. A., and N. Levinson, *Theory of Ordinary Differential Equations*, McGraw-Hill, New York, 1955.
8. Cole, R. H., *Theory of Ordinary Differential Equations*, Appleton-Century-Crofts, New York, 1968.
9. Coppel, W. A., *Stability and Asymptotic Behaviour of Differential Equations*, D. C. Heath and Company, Boston, 1965.
10. Corduneanu, C., *Almost Periodic Functions*, Interscience, New York, 1968.
11. Corduneanu, C., *Principles of Differential and Integral Equations*, Allyn and Bacon, Boston, 1971.
12. Daleckii, J. L., and M. G. Krein, *Stability of Solutions of Differential Equations in Banach Spaces*, American Mathematical Society Translations, **43**, Providence, Rhode Island, 1974.
13. Edwards, C. H., *Advanced Calculus of Several Variables*, Academic Press, New York, 1973.
14. Friedrichs, K. O., *Advanced Ordinary Differential Equations*, Gordon and Breach, New York, 1965.
15. Goldberg, J. L., and A. J. Schwartz, *Systems of Ordinary Differential Equations: An Introduction*, Harper & Row, New York, 1972.
16. Hahn, W., *Theory and Application of Liapunov's Direct Method*, Prentice-Hall, Englewood Cliffs, New Jersey, 1963.
17. Halanay, A., *Differential Equations: Stability, Oscillations, Time Lag*, Academic Press, New York, 1966.
18. Hale, J. K., *Ordinary Differential Equations*, Wiley-Interscience, New York, 1969.
19. Hartman, P., *Ordinary Differential Equations*, John Wiley & Sons, New York, 1964.
20. Hochstadt, H., Differential Equations: *A Modern Approach*, Holt, Rinehart and Winston, New York, 1964.

21. Kantorovich, L. V., and G. P. Akilov, *Functional Analysis in Normed Spaces*, Pergamon Press, New York, 1964.
22. Kartsatos, A. G., *Stability Via Tychonov's Theorem*, International Journal of Systems Science, 5, pp. 933-937, 1974.
23. Kartsatos, A. G., *Banach Space-Valued Solutions of Differential Equations Containing a Parameter*, Archive for Rational Mechanics and Analysis, 57, pp. 142-149, 1974.
24. Kartsatos, A. G., *Locally Invertible Operators and Existence Problems in Differential Systems*, Tohoku Mathematical Journal, 28, pp. 167-176, 1976.
25. Krasnosel'skii, M. A., *Translation Along Trajectories of Differential Equations*, American Mathematical Society Translations, 19, Providence, Rhode Island, 1968.
26. Ladas, G. E., and V. Lakshmikantham, *Differential Equations in Abstract Spaces*, Academic Press, New York, 1972.
27. Lakshmikantham, V., and S. Leela, *Differential and Integral Inequalities: Theory and Applications*, Vols. I, II, Academic Press, New York, 1969.
28. Liapunov, A. M., *Stability of Motion*, Academic Press, New York, 1966.
29. Lloyd, N. G., *Degree Theory*, Cambridge University Press, Cambridge, 1978.
30. McShane, E. D., *Integration*, Princeton University Press, Princeton, New Jersey, 1964.
31. Massera, J. L., and J. J. Schaffer, *Linear Differential Equations and Function Spaces*, Academic Press, New York, 1966.
32. Nemyckii, V. V., and V. V. Stepanov, *Qualitative Theory of Differential Equations*, Princeton University Press, Princeton, N. J., 1960.
33. Ostrowski, A. M., *Solution of Equations and Systems of Equations*, Academic Press, New York, 1966.
34. Pao, C. V., *On Stability of Non-linear Differential systems*, International Journal of Non-linear Mechanics, 8, pp. 219-238, 1973.
35. Reissig, R., G. Sansone, and R. Conti, *Non-linear Differential Equations of Higher Order*, Noordhoff International Publishing, Leyden, Holland,1974.
36. Roxin, E. O., *Ordinary Differential Equations*, Wadsworth, Belmont, 1972.
37. Sanchez, D. A., *Ordinary Differential Equations and Stability Theory: An Introduction*, W. H. Freeman and Co., San Francisco, 1968.
38. Schechter, M., *Principles of Functional Analysis*, Academic Press, New York, 1971.
39. Struble, R. A., *Nonlinear Differential Equations*, McGraw-Hill, New York, 1962.
40. Vidossich, G., *Applications of Topology to Analysis: On the Topological Properties of the Set of Fixed Points of Nonlinear Operators*, Nicola Zanichelli, Bologna, Italy, 1971.
41. Yoshizawa, T., *Stability Theory by Liapunov's Second Method*, The Mathematical Society of Japan, Tokyo, 1966.

INDEX

Adjoint system, 54, 101
Almost periodic function, 13
Almost periodic solution, 152
Angular distance, 144
Asymptotic equilibrium, 142
Asymptotic stability, 60, 69, 90, 128, 131
Autonomous system, 76

Banach contraction principle, 19, 119, 150
Banach space, 2
Boundary conditions, 99
Boundary value problem, 99, 109, 163
Bounded function, 12
Bounded linear operator, 7
Bounded solutions on R, 143

Cauchy-Schwarz inequality, 4
Cauchy sequence, 2
C-differentiable operator, 29
Compact operator, 26
Compact set, 22
Comparison principle, 87
Comparison theorem, 86
Continuable solution, 44

Convex set, 26

Dini derivatives, 82
Direct sum, 10

Eigenvalue of a matrix, 5
Eigenvector of a matrix, 5
ϵ-Net, 22
Equicontinuous set, 22
Equilibrium solution, 142
Equivalent norms, 3
Euclidean norm, 3
Exponential asymptotic stability, 93
Exponential dichotomy, 145
Exponential splitting, 145
Extendable solution, 44, 45, 46, 87, 89

Fixed point, 19, 109, 110, 111, 114, 135, 136, 150, 156, 158
Frechet derivative, 29, 165, 169, 170
Frechet differentiable, 29
Fredholm alternative, 102
Fundamental matrix, 51

Gronwall's inequality, 43, 53, 55, 67, 114

Index of an eigenvalue, 64
Inner product, 4
Instability, 60, 69, 97
Inverse function theorem, 31, 32, 163
Iso-stable systems, 158

Jacobian matrix, 33

Lebesgue dominated convergence theorem, 112
Leray-Schauder theorem, 27, 114, 138
Lienard's equation, 96
Linear functional, 14
Linear operator, 6
Linear system, 48
Liouville's formula, 54
Lipschitz condition, 41, 43, 49, 53
Locally invertible operator, 29
Lyapunov function, 79

Metric space, 2
Maximal and minimal solutions, 82, 83
Measure of a matrix, 65

Natural basis of R^n, 2, 3
Negatively unstable solution, 148, 149
Noncontinuable solution, 87
Nonextendable solution, 87

Normed space, 2
Norm of a Banach space, 2
Norm of a linear operator, 7
Norm of a matrix, 8
Norm of a vector in R^n, 3
Norm topology, 2

Operator, 6
Orthogonal vectors, 5
Orthonormal vectors, 5

Peano's theorem, 40
Periodic function, 12
Periodic solution, 99, 101, 135
Perturbed linear system, 70
Picard-Lindelof theorem, 41
Positive definite matrix, 5, 124
Projection matrix, 6, 10, 144

Q-bounded function, 90
Q-function, 90
Q-positive Lyapunov function, 90
Quasilinear system, 143, 156

Real normed space, 2
Region of asymptotic stability, 131, 134
Region of stability, 131
Relatively compact set, 22, 23, 25, 26

Schauder-Tychonov theorem, 26
Solution of a differential system, 40
Stability, 60, 69, 91, 128, 131, 158

Stability regions, 131
Strong stability, 60, 77
Subspace of a vector space, 1
System of differential equations, 39

T-periodic function, 12

Uniform asymptotic stability, 60, 70, 92, 128, 131

Uniform stability, 60, 69, 92, 128
Uniqueness of solutions, 41, 43, 97, 126

Van der Pol's Equation, 96
Variation of constants formula, 52

Weakly stable system, 157, 158
Weighted norm, 34, 48

SYMBOL & NOTATION INDEX

AP_n, 13

C, 5

C_ℓ^1, 168

$C_n(J)$, 12

$C_n^k(J)$, 13

C_n^{ℓ}, 13

M_n, 5

$P_n(T)$, 12

R, 1

R_+, 1

R_-, 1

$<\cdot,\cdot>$, 4

$<\cdot,\cdot>_V$, 124

$\|\cdot\|$, 2, 3, 7, 8